Laparoscopic Bariatric Surgery: Techniques and Outcomes

Eric J. DeMaria, M.D.

Rifat Latifi, M.D.

Harvey J. Sugerman, M.D.

Virginia Commonwealth University
Department of General and Trauma Surgery
Medical College of Virginia Hospitals and Physicians
Richmond, Virginia, USA

Foreword by J. Patrick O'Leary

LANDES
BIOSCIENCE
GEORGETOWN, TEXAS
U.S.A.

12-02   # 50911484

VADEMECUM
Laparoscopic Bariatric Surgery: Techniques and Outcomes
LANDES BIOSCIENCE
Georgetown, Texas U.S.A.

Copyright ©2002 Landes Bioscience
All rights reserved.
No part of this book may be reproduced or transmitted in any form or by any means, electronic or mechanical, including photocopy, recording, or any information storage and retrieval system, without permission in writing from the publisher.
Printed in the U.S.A.

Please address all inquiries to the Publisher:
Landes Bioscience, 810 S. Church Street, Georgetown, Texas, U.S.A. 78626
Phone: 512/ 863 7762; FAX: 512/ 863 0081

ISBN 1-57059-677-8

Library of Congress Cataloging-in-Publication Data

Library of Congress data applied for but not received at time of publishing.

Contents

Editors

Eric J. DeMaria, M.D.
Chapters 1, 2, 7-11

Rifat Latifi, M.D.
Chapters 1, 2 and 7

Harvey J. Sugerman, M.D.
Chapters 1, 2 and 5

Department of General and Trauma Surgery
Virginia Commonwealth University
Medical College of Virginia Hospitals and Physicians
Richmond, Virginia, USA

Contributors

Allen Donze
Stryker Endoscopy
Santa Clara, California, USA
Chapter 4

Adolfo Z. Fernandez
Department of General and Trauma Surgery
Virginia Commonwealth University
Medical College of Virginia Hospitals
  and Physicians
Richmond, Virginia, USA
Chapter 8

Giselle G. Hamad
Department of Surgery
Minimally Invasive Surgery
University of Pittsburgh School
  of Medicine
Pittsburgh, Pennsylvania, USA
Chapter 6

John M. Kellum
Department of General and Trauma Surgery
Virginia Commonwealth University
Medical College of Virginia Hospitals
  and Physicians
Richmond, Virginia, USA
Chapter 12

Shanu Kothari
Gunderson Clinic
La Crosse, Wisconsin, USA
Chapter 11

Timothy Nolan
U.S. Surgical
Norwalk, Connecticut, USA
Chapter 3

J. Patrick O'Leary
LSU Health Sciences Center
New Orleans, Louisiana, USA
Foreword

Michael Schweitzer
Division of General Surgery
Mt. Sinai Hospital
Baltimore, Maryland, USA
Chapter 9

Foreword

I was deeply honored when asked to write the foreword for this wonderful tome entitled, *Laparoscopic Bariatric Surgery: Techniques and Outcomes*. But as the glow washed over me, I immediately became concerned because of the implications of such a request. Forewords are usually either added on as an afterthought or delegated to some curmudgeon whose era of influence has passed. Either way the ego, though stroked, also becomes somewhat exposed.

As I reflect over my career in general academic surgery, I am deeply impressed with the advances that surgery has made in the treatment of individuals with serious obesity. Bariatric surgery in the beginning was at best a novelty and, in most inner circles of medicine, was condemned. Surely this condition (certainly not a disease) was simply a lack of willpower on the part of the patients so afflicted and certainly did not fall within the realm of "respectable" surgery. These patients deserved their fate because of their lack of self-control. This situation was the end result of gluttony and sloth and, if these patients had but a modicum of wit about them, they would not allow themselves to get into such "shape."

One only has to go back and look at the original publications by Payne, DeWind, and Commons to appreciate something entirely different. These California-based physicians (only one was a surgeon) saw serious obesity as a morbid condition. Not only were patients incapacitated by their excessive body weight, but they clearly identified obesity as a disease and its causal relationship with other disease states. Their publication in 1963 was, however, not the first publication related to operations for serious obesity. In the early fifties, Kremen, Linner, and Nelson at Minnesota and as a part of their work on the absorptive aspects of the jejunum and ileum, described a 385 pound patient in whom an intestinal bypass was performed. These observations were presented to the American Surgical Association Meeting in April, 1954 and, in his discussion of this paper, Philip Sandblom alluded to the fact that two years earlier Dr. Viktor Hendriksson of Sweden had performed a similar procedure on a morbidly obese patient in his homeland. Sandblom stated that the procedure had produced weight loss but that it had "created a difficult situation of nutritional balance."

Apparently, Payne, DeWind, and Commons were either aware of this previous publication or were deeply concerned about the metabolic consequences of short circuiting the intestine to the degree that they proposed. In their work they describe an extremely elaborate protocol that attempted to measure, albeit in a static state, the changes that might occur during the postoperative period. In their initial group of patients, after weight loss had occurred, they re-operated on these patients and placed their gastrointestinal tract back in continuity. They quickly learned that all of the patients so reconstructed gained back their previous weight. This greatly perplexed them. In the final three patients, they either modified the shunt or left it intact. It should be noted that only one patient in this series died and that death was apparently related to a pulmonary embolism in the late postoperative period (about six months).

It is not the intent of this discussion to outline the history of the development of bariatric surgery. Suffice it to say that a number of surgeons modified the jejunoileal bypass during the next two decades, while another cohort of surgeons attacked the problem from the other end of the gastrointestinal tract. This upstart contingency was led by Ed Mason at the University of Iowa who had actually begun his work while at Minnesota as a resident. He proposed restricting food intake by performing a procedure on the stomach that limited its reservoir capacity and shunted food into the jejunum, therefore bypassing the distal stomach, duodenum, and the most proximal jejunum. By the late 1970s, Mason had gathered together a small enclave of surgeons interested in the problem of morbid obesity. They met during June in Iowa City to discuss the developing expertise in this area. It was from this nidus that the American Society for Bariatric Surgery (ASBS) had its origins. The group was formally incorporated in the early 80s with a membership of less than 200. At the most recent annual meeting of the ASBS in excess of 1100 physicians, surgeons, and ancillary health care providers attended.

By the mid 1970s, the literature began to reflect a number of complications for the intestinal bypass procedures. Some of these were merely worrisome while others were death rendering. Concurrent to this obvious concern were benefits such as a much better understanding of the disease state of obesity, its metabolic implications, the natural history of the disease, its genetic implications, and the effects that weight loss produces on the co-morbid conditions associated with excess body weight. Surgeons also learned much about the management of seriously obese individuals who require an abdominal procedure for other causes.

Despite all of these activities, the legitimacy of surgical intervention for morbid obesity was still questioned by most of the medical community. Somehow, physicians in all specialties just couldn't accept the concept of such a major operation being performed in patients to help control excess body weight.

And then along came laparoscopic procedures. At first, approaching the abdominal cavity through a port with long instruments and a television camera just didn't seem to be true surgery. A number of individuals who were clearly leaders of surgery spoke out vehemently against the travesty of converting a relatively easy and commonly done procedure to remove the gallbladder into a video game with no one a clear cut winner. They decried the length of time the procedure took and the unacceptably high incidence of injured to adjacent organs. The parochial wisdom of surgeons stayed the course of this imaginative intervention for only a very short while. Patients considered themselves the winner. They could have such a procedure performed and found themselves shed of their diseased gallbladder with relatively little discomfort and the ability to return to work within four or five days. Patients drove the acceptance of the process and it became accepted in a very short while. In fact, if the laparoscopic approach to the procedure was not offered, most patients frequently sought another surgeon.

At this point, surgeons discovered that they can do more than remove the gallbladder using this technology. Multiple innovative techniques blossomed. Instruments improved, and finally, one daring individual decided that they would try to perform some type of gastric restrictive procedure through the scopes. At this point, as my old Irish grandmother was fond of saying, "Katie, bar the door!"

The laparoscopic gastric bypass, though performed by many early in the course, was probably brought to most surgeons' attention by Clark and Whitgrove, again from California. This operation became the tour-de-force of every laparoscopic surgeon in America and, perhaps, the world. When they had accomplished everything else, they knew they had arrived on the laparoscopic scene, if they could perform the laparoscopic gastric bypass in a patient over 325 pounds in less than five hours. In fact, many surgeons now have gotten this procedure down so that heavier patients are being operated upon with an average time of less than three hours. The patients often are discharged on postoperative day two. It has truly been miraculous!

The upside is that the technical achievements are consistent with the surgeon's abilities and the fact that, if individuals who pursue surgery as a career or challenge technically, they will rise to the occasion and accomplish many of the important adjustments necessary to reach such limits of brilliance are included in this tome. However, the downside is that patients with this disease process are not simply mannequins to test one's technical abilities. They are patients with a complex disease process, and the operations that we perform produce a state of metabolic jeopardy that must be managed over a protracted period of time. In the beginning, surgeons took on the commitment of follow-up for the patient's life span. I do not see any reason to abandon that initial tenet at this stage. Simply because we can do the operations with greater facility and have patients out of the hospital quicker does not mean that the procedure has any less risk in the late postoperative period. Patients are still at risk for metabolic derangement years after the procedure and need to be followed, not only for such abnormalities.

We must also use ancillary healthcare professionals to support patient's psychiatric adjustment to their new-found involvement in society.

Morbid obesity is a serious disease that is multifactorial. It produces a state of social maladaptation, compromised socioeconomic state, and a vulnerability that is infrequently recognized. Patients operated upon experience enormous changes in their life, as well as in their metabolic state and need to be followed for a protracted period of time. This tome deals in great detail with the procedures and approach to morbidly obese patients. It will be a great asset to all surgeons involved in the care of these patients. However, technical success, if not coupled with care of the total patient, is a hollow victory.

J. Patrick O'Leary, M.D.
The Isidore Cohn, Jr. Professor
and Chairman of Surgery
LSU Health Sciences Center
New Orleans, Louisiana, U.S.A.

CHAPTER 1

Introduction

Bariatric surgery has finally been recognized as the only treatment for morbid obesity with successful outcomes in long-term follow-up. Although the ultimate solution for treatment of this disease may involve metabolic or genetic manipulations, currently no such treatments generate weight loss results and improvements in medical co-morbid conditions which compare to surgical treatment. However, despite successful outcomes and published low complication rates in major centers, bariatric surgery is often viewed by the public and prospective patients as a drastic, invasive, and high risk solution. The American public does not uniformly embrace the surgical treatment option for morbidly obese patients. Letters to *People* magazine, even in the aftermath of celebrity Carnie Wilson's highly successful gastric bypass surgery, sadly reveal the altogether too frequent prejudice and innuendo about laziness, sloth, and below average intelligence that sufferers of morbid obesity must endure as once again the public demonstrates that they do not understand obesity and are unsympathetic to its victims.

In contrast to the stereotypes of morbidly obese patients, bariatric surgeons recognize that there is no more rewarding area of surgical specialization than treatment of the morbidly obese patient. Obese patients are often highly motivated to reclaim a more normal life and eliminate their co-morbid medical problems. Our co-editor Harvey Sugerman has expressed it best: "Bariatric surgery is like no other treatment in medicine. There is no other situation in which a surgeon can perform a single surgical procedure and cure 4 or 5 of the patient's medical problems." Despite the challenges and pitfalls of caring for obese patients, a successful surgical outcome usually leads to an extremely grateful patient and family.

The introduction of minimally invasive surgical approaches to treat obesity has led to a tremendous interest in bariatric surgery in the United States and around the world. Laparoscopic surgery represents a tremendous opportunity to ameliorate the public's fear that surgical treatment is too drastic and invasive to be considered a reasonable choice for the estimated 12 million Americans who could benefit greatly from successful weight loss. This challenging new area of specialization within bariatric surgery represents a dramatic opportunity to help so many, yet it is fraught with myriad obstacles for safe development. There is no more technically demanding laparoscopic procedure being offered on a routine basis today than laparoscopic Roux-en-Y proximal gastric bypass. Furthermore, training and skill in advanced laparoscopy are an inadequate background for developing a program since the bariatric patient presents many complex issues in management for the surgeon. Thus, the novice must master the skills of two disciplines: advanced laparoscopy and bariatric surgical care.

Laparoscopic Bariatric Surgery: Techniques and Outcomes, edited by Eric J. DeMaria, Rifat Latifi and Harvey J. Sugerman. ©2002 Landes Bioscience.

1

This text provides a resource to surgeons committed to bringing safe and effective laparoscopic treatment to their morbidly obese patients. Its rapid publication addresses the usual concerns that textbooks lag significantly behind developments in the field, a phenomenon which would have been totally unacceptable in this rapidly emerging field. Each chapter addresses issues which are critical to the safe application of minimally invasive techniques to the unique population of morbidly obese patients. The authors bring forward their experience in bariatric surgery and lessons learned from the development of their laparoscopic obesity surgery program.

The Editors

CHAPTER 2

Indications and Patient Selection for Bariatric Surgery

Rifat Latifi, Eric J. DeMaria and Harvey J. Sugerman

Introduction

Approximately 97 million adults in the United States are overweight or obese; 32.6 % are overweight, defined as a body mass index (BMI) of 25-29.9 kg/m², while 22.3% are obese with a BMI >30 kg/m². Morbid obesity, or clinically severe obesity is defined as 100 lb. above ideal body weight, or a BMI ≥35 kg/m². Severe obesity (more than 244 lb. for men or more than 225 lb. for women) has been estimated to be present in 4.9% (2.8 million) of men and 7.2% (4.5 million) of women in the United States. As the BMI increases, so does the mortality rate from all causes, especially from cardiovascular disease, which is 50%-100% above that of persons who have BMI in the range 20-25 kg/m².

Pathophysiology of Morbid Obesity Syndrome

Morbid obesity is a potentially deadly syndrome that is a harbinger of multiple other diseases and disorders, affecting every organ and system of the body, and as such it is associated with several significant clinical conditions.

Cardiovascular Related Problems

Cardiovascular dysfunction in morbidly obese patient is common and is manifested as hypertension, coronary artery disease, and increased complications following coronary artery bypass et cetera. Heart failure may be the consequence of left ventricular hypertrophy and hypertension, left ventricular eccentric hypertrophy, or right ventricular hypertrophy. In addition prolonged Q-T interval and sudden death occur more commonly in morbidly obese patient. Furthermore, patients with hypoventilation syndrome have higher cardiac filling (pulmonary artery and pulmonary capillary wedge) pressures that are higher than pressures in patients with congestive heart failure (CHF), although clinically they are not in CHF. Other problems such as dysrhythmias, ischemic stroke, deep vein thrombosis and pulmonary embolus are common in-patients with morbid obesity.

Respiratory Insufficiency

Respiratory insufficiency of obesity (Pickwickian syndrome) is associated with obesity hypoventilation syndrome and obstructive sleep apnea syndrome (multiple nocturnal awakenings, loud snoring, falling asleep while driving, daytime somnolence). In addition to high filling pressure, most of these patients have abnormal pulmonary function tests.

Laparoscopic Bariatric Surgery: Techniques and Outcomes, edited by Eric J. DeMaria, Rifat Latifi and Harvey J. Sugerman. ©2002 Landes Bioscience.

2

Metabolic Complications

The relationship of central obesity to the constellation of health problems known as "the metabolic syndrome" or "syndrome X" is well established. It is thought at the present that increased visceral fat increases glucose production that subsequently causes hyperinsulinism and eventually development of type II diabetes mellitus. It has been also established that increased glucose and insulin levels are responsible for polycystic ovary syndrome (Stein-Leventhal syndrome) with ovarian cysts, hirsutism, and amenorrhea as well as non-alcoholic steatohepatitis (NASH) which may progress to liver cirrhosis. Type II diabetes mellitus can cause significant morbidity as it may lead to renal failure, peripheral neuropathy and retinopathy. Other metabolic complications are hypertension, elevated triglycerides, cholesterol and in increased frequency of formation of gallstones.

Increased Intra-Abdominal Pressure

Increased intra-abdominal pressure is well documented in morbidly obese patients and is manifested as stress overflow urinary incontinence, gastroesophageal reflux, nephrotic syndrome, increased intracranial pressure leading to pseudotumor cerebri, hernias, venous stasis, probably hypertension and pre-eclampsia, as well as the nephrotic syndrome. Pseudotumor cerebri is hypothesized to be secondary to an increased intra-abdominal pressure and intra-thoracic pressure with decreased venous drainage from the brain.

Other Co-Morbid Conditions

Hypercoagulapathy, female hormonal dysfunction such as amenorhea, dysmenorrhea, infertility, hypermenorrhea, increased incidence of breast cancer, uterine, colon, prostate and other cancers and debilitating joint disease involving hips, knees, ankles, feet and lower back are common in these patients. In addition, obese patients clearly experience a multitude of difficulties related to social acceptance in society, work-related problems, body image, reduced mobility, sexual dysfunction and other psychosocial problems that add more pathology to this chronic and deadly disorder. The difficulties in diagnosing and treating surgical conditions in obese patients such as peritonitis, necrotizing panniculitis, necrotizing fasciitis, diverticulitis, necrotizing pancreatitis and other intra-abdominal infectious catastrophes are significant.

A relationship between obesity and a low-grade systemic inflammatory state has been established and is manifested with elevated proinflammatory cytokines interleukin 6 (IL-6) and C-reactive protein which are thought to contribute to cardiovascular morbidity. Elevated C-reactive protein levels have been associated with increased risk of myocardial infarction, stroke, peripheral arterial disease, and coronary heart disease.

Indications for Surgical Treatment of Morbid Obesity

The causes of severe obesity are multifactorial. However, the chronic imbalance between energy intake and energy expenditure is the most common cause. Genetic and other environmental factors play an important role too.

The published success rate for all medical approaches including diet, pharmacotherapy and behavioral modification for morbid obesity is dysmal. It has been estimated that over 95% of morbidly obese patients subjected to medical weight-reduction programs regain all of their lost weight, as well as additional excess weight, within two years of the onset of therapy.

Because of the very high failure rate of all non-surgical attempts to correct morbid obesity including diet, behavior modification, hypnosis, voluntary incarceration, jaw wiring and intragastric balloons, the presence of morbid obesity by itself may be an indication for surgical correction. Based on medical evidence, the surgical treatment of patients with BMI >40 kg/m^2 or BMI> 35 kg/m^2 with co-morbid conditions, has emerged as definitive therapy. Recently, bariatric surgery has gained acceptance among surgeons, physicians and the lay public. The presence of any endocrine disorder that may be responsible for obesity, albeit extremely rare, should be treated first.

Preoperative Patient Evaluation

One of the most important factors in the success of bariatric surgery is appropriate selection of patients. It is clear that surgery is the best method thus far to induce significant weight loss in morbidly obese patients. Yet at the same time the patient should understand that surgery alone is not enough, and that this procedure requires life-long commitment of the patient, their family and their physicians. As recommended by the NIH panel, a patient is a candidate for surgery if he/she has failed nonoperative attempts to lose weight, is motivated and is not addicted to drugs or alcohol. The patient must understand the magnitude of their problem and be able follow the postoperative dietary and other regimens.

Morbidly obese patients requesting bariatric surgery need to be evaluated by their primary medical doctor and bariatric surgical team that consists of a dedicated and experienced surgeon, dietitian, and a nurse. A psychologist and or psychiatrist specializing in behavior modifications should be available for consultation. A very frank conversation(s) between the surgeon and patient and family is most important.

Patients need to understand clearly all phases of the treatment, the procedure itself, and possible complications. While it is a very important step, the surgery itself is probably the easiest phase of what will be lifetime changes for these patients. Only when the patient understands all the intricacies of the procedure, the course of postoperative care and is deemed to be a proper candidate for surgical treatment, should one proceed with surgery. In today's information age when the procedure may be seen on the internet, many patients are much more sophisticated and informed about the operation, the existing expertise, method and techniques of the operation. As part of the preoperative evaluation, the surgeon should perform a careful physical examination and take a complete medical history. Most of the patients have co-morbid diseases and those need to be sought out carefully. In our practice, the surgical team is often the first to suspect the diagnosis of sleep apnea syndrome. Furthermore, the dietary habits, social situation and motivation for the operation should be questioned as well as the history of obesity in the family. Basic laboratory work-up should include a complete blood count, full chemistry panel, iron, B12, thyroid panel, arterial blood gases on room air, EKG, a chest X-ray and urinalysis. If history indicates the suspicion for existence of sleep apnea, the patient should undergo a sleep study. Many morbidly obese patients are seen for the first time after they are involved in trauma situation after falling asleep while driving. These patients should

2

Table 2.1. Criteria for patient selection for bariatric surgery*

- Weight
- 45 kg or 100% above desirable weight
- Body mass index>40 kg/m²
- Body mass index >35 kg/m² with coexisting morbidities
- Failure of non-surgical methods of weight reduction
- Absence of endocrine disorders responsible for morbid obesity
- Psychological stability
- Lack of drugs and alcohol abuse

---

*The 1991 National Institutes of Health Consensus

---

be referred to the bariatric surgeon by the trauma team. The history and physical examination should identify the patients who need further work-up from a cardio-vascular standpoint, such as those with hypoventilation syndrome or venous disease. While the choice of bariatric procedure recommended to the patient is based mainly on the local expertise and the tradition, we recommend mostly Roux-en-Y gastric bypass (RYGBP). Whether this procedure is performed laparoscopically or by an open technique, depends on the individual surgeon and patient request. Previous abdominal operations may be a relative contraindication to laparoscopy especially if the surgery was limited to upper abdomen, such as gallbladder surgery. The presence of hypoventilation syndrome is not a contraindication for laparoscopic surgery, especially when the surgeon and the anesthesia team monitor the patient carefully. While some had identified a large left lobe of the liver as a difficult problem, this is not a contraindication to laparoscopic gastric by-pass surgery, although it may be present a significant problem.

Effects of Weight Loss: What Can the Patient Expect?

There are many studies which document the reversibility of most co-morbid obesity conditions. Type II diabetes mellitus will resolve in 85% of patients so that they no longer require insulin or any oral hypoglycemic medication. Obstructive sleep apnea syndrome resolves completely when the respiratory disturbance index is <40 episodes/hour of sleep and improves significantly if >40 at 5 years after surgery. In addition, obesity hypoventilation resolves within 6 to 9 months after surgically induced weight loss with improvement in arterial blood gases, lung volumes and cardiac filling pressures. Cardiac function improves significantly following surgically induced weight loss. Systemic hypertension resolves in two-thirds to three-fourths of the patients who no longer need any anti-hypertensive medications or respond to a much smaller dose. There is a marked improvement in serum lipids following gastric bypass, as well as correction of urinary overflow incontinence in women. Gastroesophageal reflux disease (GERD) is no longer a problem in almost all patients immediately after GBP surgery as there is no acid or bile to reflux from the small gastric pouch; however, this can be a serious complication of vertical banded gastroplasty (VBG) necessitating conversion to GBP. Venous stasis ulcers will heal and lower extremity peripheral edema resolve following surgically induced weight loss, presumably as a result of decreased abdominal pressure on the inferior vena cava. Pregnancy may be a complication of bariatric surgery as women begin to ovulate and become fertile; unfortunately, their hirsutism will not go away with weight loss. It is recommended that women take contraceptive precautions for 1 year after bariatric surgery because of the potential risk of neural tube defects (spina bifida, etc.) with nutritional impairment during pregnancy. Pseudotumor cerebri has also been shown to resolve after surgically induced weight loss; these patients no longer suffer from constant headaches and pulsatile tinnitus and their opening cerebrospinal fluid (CSF) pressures normalize. Patients with degenerative joint disease involving the hips, knees, ankles and lower back will usually claim a marked decrease in pain and improved mobility following marked weight loss, but there are no studies documenting this impression to date. It may obviate the need, either temporarily or longer, for artificial joint replacement. There are no data to date evaluating the effect of major weight loss on non-alcoholic steatohepatitis (NASH), although the standard recommendation for these patients is to lose weight. Psychological evaluation has found a significant improvement in self-image and symptoms of depression, but one study has noted that the severity of depression may return to pre-surgical levels at 5 years after surgery in the absence of weight regain and may be associated with the risk of suicide.

The Reason(s) for Inadequate Weight Loss or Weight Regain Following Gastric Bypass Surgery

The average patient loses two-thirds of their excess weight, or about one-third of their preoperative weight, following a gastric bypass procedure. At five years after surgery, the average loss of excess weight is 60% and it is 50% at ten years. Better long-term weight loss may be seen following the partial biliopancreatic bypass or duodenal switch malabsorptive procedures although at the potential cost of malnutrition. Approximately 15% of patients will fail to lose more than 40% of their excess weight following a gastric bypass procedure. This percentage is much higher following banded gastroplasty procedures and, presumably, following laparoscopic gastric banding. In patients who have undergone either a stapled gastroplasty or gastric bypass, there is always a possibility of staple line disruption, especially if the patient states they are able to eat much larger quantities of food at a time. An UGI should determine if this complication has occurred and, should that be found, operative revision can be undertaken. Revisional procedures in bariatric surgery are associated with a higher frequency of complications, including anastomotic leak. The primary cause of failed weight loss following gastric bypass is the frequent ingestion of high fat junk (potato or corn chips) and fried foods (French fried potatoes) or the ability to tolerate high-density carbohydrates such as non-dietetic sodas, lemonade, cookies and ice cream. Dilation of the gastrojejunal stoma does occur but surgical revision does not lead to weight loss. Conversion to a malabsorptive distal gastric bypass is effective for improved weight loss but risks the development of protein-calorie malnutrition and steatorrhea with foul-smelling stools and fat-soluble vitamin deficiencies. It is thought that only patients with severe obesity co-morbidity (severe hypertension resistant to drug therapy, obesity hypoventilation or diabetes) who have failed a standard gastric bypass should be offered conversion to a malabsorptive procedure and then only after thoroughly informed consent as to its risks.

Selected Readings

1.   NIH Conference: Methods for voluntary weight loss and control. NIH Technology Assessment Conference Panel. Consensus Development Conference, 30 March to 1 April 1992. Ann Intern Med 1993; 119:764-70.

2.   NIH Conference: Gastrointestinal surgery for severe obesity: Consensus Development Conference Panel. Ann Intern Med 1991; 115:956-61.

3.   Johnson D, Drenick EJ. Therapeutic fasting in morbid obesity. Arch Intern Med 1977; 137:1381-2.

4.   Sugerman HJ, Kellum JM JR, DeMaria EJ, Reines HD. Conversion of failed or complicated vertical banded gastroplasty to gastric bypass in morbid obesity. Am J Surg 1996; 171:263-9.

5.   Sugerman HJ, Starkey JV, Birkenhauer RA. A randomized prospective trial of gastric bypass versus vertical banded gastroplasty and their effects on sweets versus non-sweet eaters. Ann Surg 1987; 205:613-24.

6.   Hall JC, Watts JM, O'Brien PE, et al. Gastric surgery for morbid obesity. The Adelaide study. Ann Surg 1990; 211:419-27.

7.   Schauer PR, Ikramuddin S, Ramanathan R, Gourash W, Panzak G. Outcomes after laparoscopic Roux-en-Y gastric bypass for morbid obesity. Ann Surg 2000; 232:515-29.

8.   Pories WJ, Swanson MS, Macdonald KG. Long SV, et al. Who would have thought it? An operation proves to be the most effective therapy for adult-onset diabetes mellitus. Ann Surg 1995; 222:339-50.

9.   Sugerman HJ, Fairman RP, Sood RK, Engle K, Wolfe L, Kellum JM. Long-term effects of gastric surgery for treating respiratory insufficiency of obesity. Am J Clin Nutr 1992; 55:597S-601S.

10. Sugerman HJ, Windsor ACJ, Bessos MK, Wolfe L. Abdominal pressure, sagittal abdominal diameter and obesity co-morbidity. J Int Med 1997; 241: 71-9.

11. Sugerman HJ, Baron PL, Fairman RP, Evans CR, Vetrovek GW. Hemodynamic dysfunction in obesity hypoventilation syndrome and the effects of treatment with surgically induced weight loss. Ann Surg 1988; 207:604-13.

12. Sugerman HJ, Brewer WH, Shiffman ML, et al. A multi-center, placebo-controlled, randomized, double blind, prospective trial of prophylactic ursodiol for the prevention of gallstone formation following gastric-bypass-induced rapid weight loss. Am J Surg 1995; 169:91-7.

13. Sugerman HJ, Kellum JM, Reines HD, et al. Greater risk of incisional hernia with morbidly obese than steroid dependent patients and low recurrence with pre-fascial polypropylene mesh. Am J Surg 1996; 17:80-4.

14. Sugerman HJ, Felton WL, Sismanis A, Salvant JB, Kellum JM. Effects of surgically induced weight loss on pseudotumor cerebri in morbid obesity. Neurology 1995; 45:1655-9.

15. Sugerman HJ, Felton WL III, Sismanis A, Kellum JM, DeMaria EJ, Sugerman EL. Gastric Surgery for pseudotumor cerebri associated with severe obesity. Ann Surg 1999; 229:634-42.

16. Sugerman HJ, DeMaria EJ, Felton WL III, et al. Increased intra-abdominal pressure and cardiac filling pressures in obesity-associated pseudotumor cerebri. Neurology 1997; 49:507-511.

CHAPTER 3

Laparoscopic Instruments for Bariatric Surgery

Timothy J. Nolan

Introduction

Laparoscopic gastric bypass (LGB) has developed in recent years due to close collaboration between physicians and the technology manufacturers. Close relationships with surgeons have allowed our industry to recognize the emergence of this procedure, and start to develop technologies that help make a LGB a safe, effective option for a challenging patient population.

This chapter will discuss the instruments and technology currently used by many surgeons to perform LGB. The examples presented are from United States Surgical* but many are available from other manufacturing sources. As this writing is about a most advanced and challenging procedure, it is assumed the reader has an understanding of basic laparoscopic equipment. It is important to recognize that although current technology has enabled surgeons to perform LGB safely and effectively, further advances are needed to aid surgeons in their continued efforts to refine this procedure.

Laparoscopic Access

Initial access for insufflation can be achieved by the use of an insufflation needle or through open technique. For the bariatric patient, the Surgineedle™ is available in a longer length (150mm) (Fig. 3.1). This single use pneumoperitoneum needle is designed to help protect internal viscera from inadvertent injury; it has a blunt stylet that advances over the sharp tip upon entry to the abdominal cavity. For open access, the Blunt Tip Trocar (Fig. 3.2) provides an airtight seal by inflating the distal balloon against the peritoneum and securing a soft foam anchor against the skin. This eliminates the need for fascial sutures (no easy task in the morbidly obese) and is more forgiving to variations in open cut down techniques.

As LGB is a most technically challenging procedure, high quality operative ports are essential for the surgeon to maintain concentration on the task at hand. Versaport™ V² (Fig. 3.3) allow the surgeon to exchange instruments that are 5mm in diameter up to the trocar's maximum without stopping to place converters on the valve system. Morbidly obese patient often have extremely large livers, and excessive intra-abdominal fat and single use, shielded trocars are most often used. The Versaport™ V² trocar has a shielded linear blade designed to help the surgeon achieve a controlled safe entry into the abdominal cavity. When trocar anchors, such as Surgigrips™ (Fig. 3.4) are used, it is important to secure them into the fascia to hold securely. This may be challenging in large patients, and surgeons often utilize

Laparoscopic Bariatric Surgery: Techniques and Outcomes, edited by Eric J. DeMaria, Rifat Latifi and Harvey J. Sugerman. ©2002 Landes Bioscience.

3

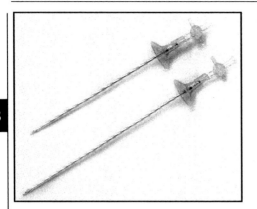

Fig. 3.1. Surgineedles™

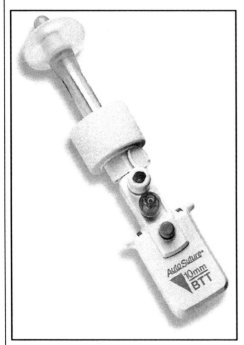

Fig. 3.2. Blunt Tip Trocar

sutures to additionally secure the ports to the skin. As with all laparoscopic procedures, the facial defect from trocar sites, 10mm and larger, are commonly closed
with suture. Suture passing instruments such as the Endo Close are commonly
used (Fig. 3.5).

An alternative to traditional trocars, the STEP™ access system (Fig. 3.6) allows
the surgeon to create intra peritoneal access without using cutting trocars. Radial
dilation technology is used to expand a 14 gauge insufflation needle up to a 12mm

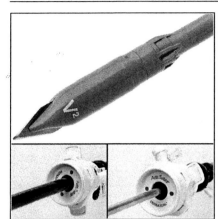

Fig. 3.3. Versaport™ V² trocars.

3

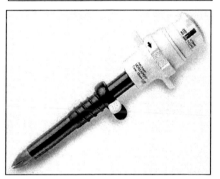

Fig. 3.4. Surgigrips™

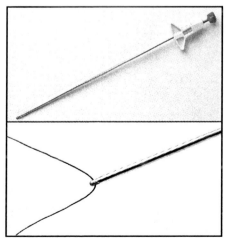

Fig. 3.5. Endo Close™ suture passing needle.

3

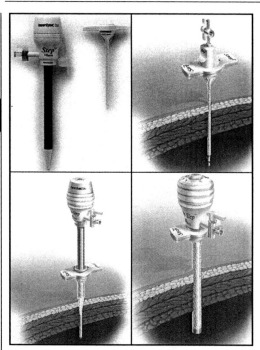

Fig. 3.6. STEP™ access system.

working port. A 12mm port radially dilates the entry and leaves a residual defect of only 6 mm. Most surgeons feel this small residual defect eliminates the need to close the port site with suture, which can be quite challenging in obese patients. Radially dilated access also secures the port tightly within the abdominal wall. This greatly reduces the possibility of the port inadvertently pulling out from the abdominal wall during instrument exchange. Many surgeons find this a true benefit as reinserting ports in morbidly obese patients can be difficult and raises the potential for creating additional facial defects. The new VersaStep™ access system, combines the advantages of radial dilation access with the converterless VersaSeal™ system. It is available in standard and extra long cannula lengths to accommodate variations in abdominal wall thickness (Fig. 3.7).

Hand Instruments—Retractors, Graspers, Dissector and Scissors

As stated earlier, morbidly obese patients often have very large and heavy livers. Good liver retractors are essential in gaining the exposure required to create the gastric pouch.

One example is the Endo Paddle Retract ™ (Fig. 3.8). Fully deployed, the wide face covers a large surface area reducing the likelihood of the liver slipping away.

It has become clearly known that the most important tools for this procedure are a traumatic graspers: four atraumatic graspers (2 for the surgeon and 2 for the assistant), back up atraumatic graspers, and an extra set of atraumatic graspers. There is

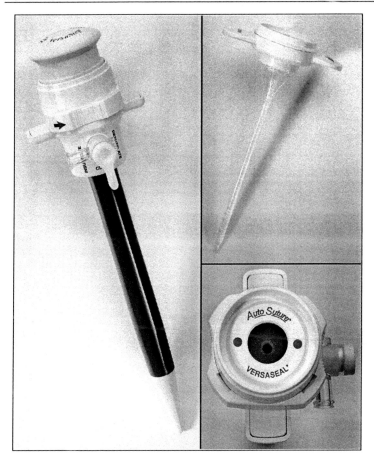

Fig. 3.7. The VersaStep™ access system.

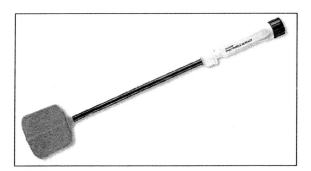

Fig. 3.8. Endo Paddle Retract™.

3

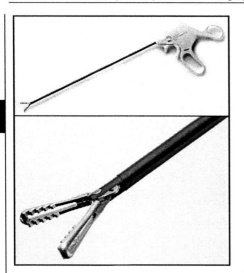

Fig. 3.9. Endo Clinch™.

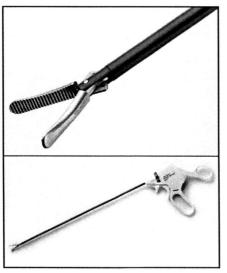

Fig. 3.10. Endo Grasp™.

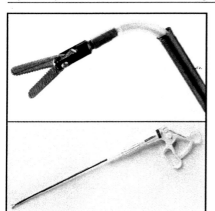

Fig. 3.11. Roticulator™.

3

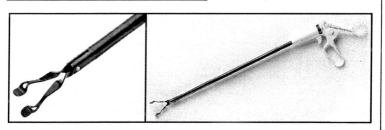

Fig. 3.12. Endo Babcock™.

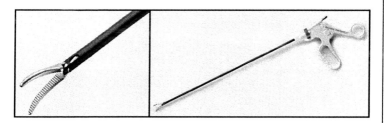

Fig. 3.13. Endo Dissect™.

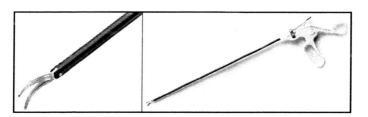

Fig. 3.14. Endo Shears™.

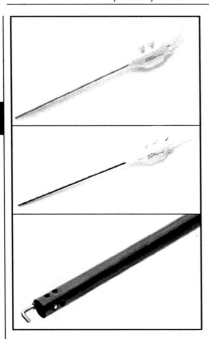

Fig. 3.15. Surgiwand™.

3

an extensive and varied amount of visceral manipulation required in LGB, from retracting the stomach to running the small bowel. "Standard" laparoscpic instrument trays are not likely to meet a surgeon's needs for LGB. Reusable and single use hand instruments are available from many sources. The following instruments (Figs. 3.8-3.13), from U.S.S., are common choices for LGB and are representative of the designs most commonly used (Endo Clinch™, Endo Grasp™, Endo Babcock™, Roticulator™, Endo Dissect™, Endo Shears™).

Suction/Irrigation

As with any laparoscopic procedure, a suction/irrigation device is used to keep the surgical field clear of pooled blood. Due to the extensive amount of dissection required in LGB, suction is often used to evacuate cautery smoke and ultrasonic vapor. The Surgiwand™ suction/irrigation device (Fig. 3.15) utilizes a standard trumpet valve to provide both suction and irrigation through a common 5mm channel. Integral hook and spatula cautery are available.

Clip Appliers

Surgeons performing LGB believe automatic laparoscopic clip appliers are essential, not as much for predetermined tasks as managing unexpected bleeding. As they are routinely used in this fashion, the most important attribute of a clip applier is that it is readily available in the operating theatre. Even though this might seem obvious, it is still important to mention. Recently, 5mm clip appliers have become very popular in laparoscopy and are ideal for LGB when 5mm ports are used. The

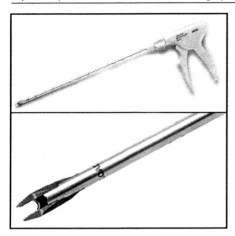

Fig. 3.16. Endo Clip™ 5mm automatic clip applier.

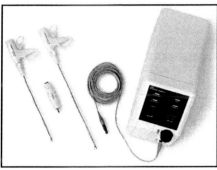

Fig. 3.17. AutoSonix™ Ultra Shears™ ultrasonic coagulating device.

ability to use the Endo Clip™ 5mm automatic clip applier (Fig. 3.16) from any port location is advantageous in controlling unexpected bleeding.

Ultrasonic Coagulation

In recent years, ultrasonic coagulation technology has played an important role in the advancement of laparoscopic general surgery. Ultrasonic coagulation devices allows for dissection of tissues and coagulation of small vessels without causing thermal damage to surrounding tissues. This has allowed surgeons to perform complex laparoscopic dissections with confidence. Smoke created by electro-surgical devices can interfere with the surgeon's tasks and is troublesome to evacuate. Ultrasonic Coagulation devices create a limited amount of vapor, and thus do not impede the progress of the surgeon. Dissection near the gastro-esophageal junction can be difficult with standard devices. The AutoSonix™ UltraShears™ ultrasonic coagulating device is available with a longer, 37 mm shaft, specifically designed for LGB (Fig. 3.17). This added length greatly facilitates use of the instrument in morbidly obese patients.

3

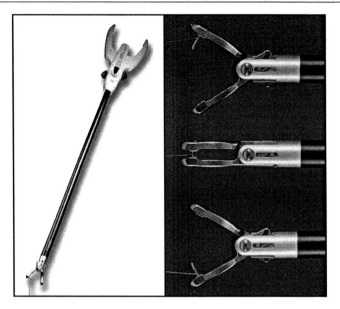

Fig. 3.18. Endo Stitch™ suturing device.

Laparoscopic Suturing Devices

Laparoscopic suturing is an essential part of performing LGB. Sutures are used for closing mesenteric defects; stay sutures may be needed for stapled anastomoses. Sutures are sometimes used to reinforce or close stapled anastomoses, and some surgeons even perform completely sutured anastomoses. High quality laparoscopic needle drivers are available from manufacturers of high quality hand instruments. Many surgeons prefer an automated suturing device. The Endo Stitch™ suturing device (Fig. 3.18) allows surgeons to perform laparoscopic suturing tasks quickly and effectively. The two essential components of endoscopic suturing, needle passing and knot tying, can both be accomplished with this device. This 10mm device can be used with both absorbable and non-absorbable braided sutures. The suture is wedged between the two taper points of the needle that is loaded into the jaws of the instrument. The needle shuttles from one jaw to the other as it passes through tissue. This eliminates the need for repositioning and the possibility of a loose needle falling from laparoscopic view. Knot tying techniques for both continuous and interrupted suturing are easily mastered.

Endoscopic Stapling Devices

For the past 30 years, surgical staplers have been standard tools for transection and creation of anastomosis in gastro-intestinal surgery. The introduction of laparoscopic staplers in the early 1990s allowed surgeons to perform standard procedures with minimal access. A primary function of early laparoscopic staplers like the Endo GIA™ 30mm stapler was for vascular applications in general surgical procedures such as and solid organ removal. It can be argued that advances in laparoscopic

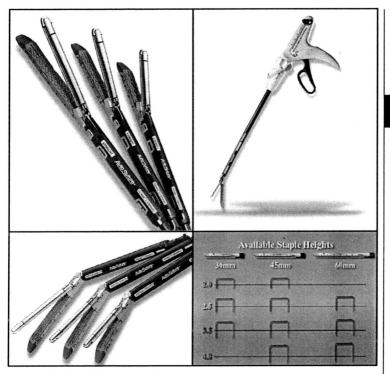

Fig. 3.19. Endo GIA™ Universal Stapling System.

surgical staplers have been the technological driving force in LGB. An important advance in this area has been the introduction of smaller staples, 2.0mm, for many vascular applications. The vascular staplers with 2.5mm staples were originally designed for large dissected vessels such as pulmonary vessels. The need for vascular staples in LGB is far different. Vessels encountered in the division of small bowel mesentery and in mobilizing the stomach are small by comparison to pulmonary vessels. As a result 2.0mm staples were developed for these applications.

In the 1990s, only a limited number of surgeons used Laparoscopic staplers to perform intra-corporal gastro-intestinal anastomosis. They had developed techniques that often required multiple intra-luminal applications and did not have the ease of open surgery. The Endo GIA™ Universal stapling system, utilizes a completely new technology and allows surgeons to perform laparoscopic transection and anastomosis using stapling techniques similar to open surgery. A cost effective and practical device, this stapling system fires fixed or Roticulator™ loading units of three linear lengths, 30mm, 45mm, and 60mm (Fig. 3.19). For the needs of laparoscopic bariatric surgery, the Endo GIA Universal™ XL is ten centimeters longer than the standard stapler. This enables better access to the surgical site in patients with extremely thick abdominal walls (Fig. 3.20). Earlier 60mm laparoscopic linear cutting staplers

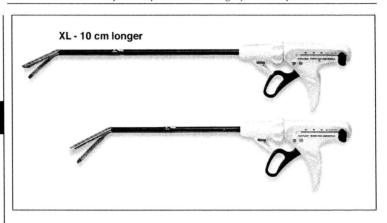

Fig. 3.20. Endo GIA Universal™ XL and Endo GIA Universal.

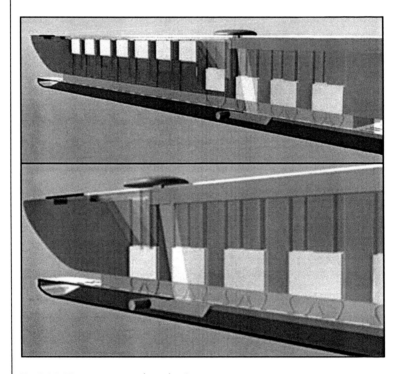

Fig. 3.21. Tissue gap control mechanism.

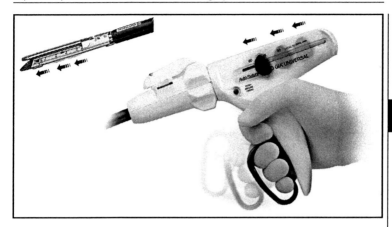

Fig. 3.22. Firing concept similar to low gearing concept.

were not embraced by most surgeons, as they were cumbersome to use, and required trocars up to 18mm in diameter. This was due to the limits of standard tissue compression technology. The diameter of the clamping mechanism was proportional to the linear length of the staple line. The Endo GIA™ Universal utilized a unique tissue compression technology. To avoid the need for a large aggressive clamping mechanism, a tissue gap control mechanism, similar to an I-beam, is employed (Fig. 3.21). This mechanism physically connects the staple cartridge and anvil. As the instrument is fired, the tissue gap mechanism gradually compresses the tissues and brings the staple cartridge and anvil into proper apposition for precise staple formation along the entire length of the staple line. To reduce excessive force when firing on thick tissue, the firing mechanism gains a mechanical advantage using a low gearing concept. This simple concept is similar to the lower gearing of a bicycle, which makes climbing a steep incline possible. (Fig. 3.22).

Some surgeons utilize a circular stapler for the creation of the gastro-jejunal anastomosis in gastric bypass surgery. Several modifications to standard devices have been made to facilitate laparoscopic use. The Premium Plus Curved EEA™ stapler (Fig. 3.23) is an example of such a device. Seals have been added to prevent leaking of CO_2 though the instrument body. To improve access to the operative site in laparoscopic procedures, a streamlined body was designed, increasing the working length of the instrument. A suture hole was added to the removable trocar to facilitate removal from the instrument and withdrawal from an access port. An additional suture hole was added to the center rod of the anvil. This enables the surgeon to guide the anvil into position utilizing a suture lead. Additional modifications facilitate mating the instrument and anvil during laparoscopy. A grasping notch was added to the center rod allowing the surgeon to control the anvil with standard laparoscopic forceps. Obtaining the proper angles to mate the anvil and stapler head is challenging in laparoscopy due to the fixed perspective of the access ports. A new geometry to the center rod has been employed to aid the surgeon in this task.

3

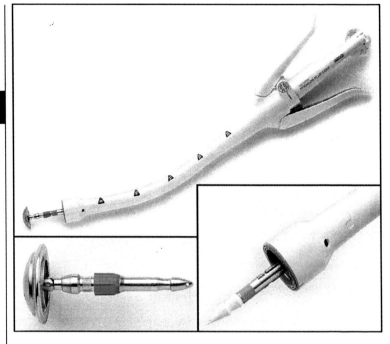

Fig. 3.23. Premuim Plus Curved EEA™ stapler.

In summary, the devices and equipment described represent the latest advances in laparoscopic instrumentation and are used in wide range of advanced procedures. Ongoing development of instrumentation specific to morbidly obese patients is needed. Industry must work side by side with the surgical community provide the technologies that will enable the laparoscopic bariatric surgeon to continually improve outcomes and best serve this important patient population.

CHAPTER 4

State of the Art Endosuite® for Minimal Access Surgery

Alan Donze

Introduction

Over the last nine years there has been a great movement towards designing a more perfect surgical operating theatre and environment for the demanding requirements of advancing minimally invasive surgical procedures. Evidence of this movement is abundant in the over 600 surgical Endosuites® Operating Rooms (OR) installed throughout the United States since 1992. The purpose of the following section is to provide information regarding the underlying technologies, equipment, and design philosophies behind today's operating room of the future.

Basic Philosophy of the MIS Endosuite™

Until the early 1990s, operating rooms were constructed much the same as they had been for nearly 100 years. In the late 1880s, Dr. Charles McBurney designed the William J. Sym's Operating Pavilion, which proved to be the standard for 20th century operating rooms. The suite emphasized aseptic design—cleanable nonporous surfaces on walls and flooring, minimal furniture, specially constructed shelves and tables of metal and glass, and heat sterilization. The room was rectangular in shape with a special operating table in the center of the room.

As minimally invasive surgical procedures increased in the early 1990s, patients benefited from significantly reduced recovery times. However, the OR staff was hurled into a new age of technology, with little or no preparation, leading to significant procedural down time and inconsistent delivery of the appropriate instrumentation, equipment, and ultimately patient care.

Surgeons were unhappy as a result of significant increases in turnover time and procedure down time. At the same time, managed care was forcing a reduction of OR staffing levels nationwide. Adding to the problem was an increasingly cluttered OR environment due to the addition of endoscopic video equipment towers and other equipment. This equipment often proved to be complicated to use and expensive to repair.

With increasing costs and time, decreasing costs and personnel, and an environment that promoted confusion rather than patient care, the operating room as designed by Dr. McBurney had been stretched beyond its capabilities.

In 1993, Stryker Endoscopy began marketing the Endosuite® Operating Room— OR of the Future concept. The basis of this concept is to address the inefficiencies of most operating rooms in use today such as inefficient use of time, space, information and personnel. In addition to the inability to efficiently control

Laparoscopic Bariatric Surgery: Techniques and Outcomes, edited by Eric J. DeMaria, Rifat Latifi and Harvey J. Sugerman. ©2002 Landes Bioscience.

equipment and the environment. Thus, today's minimal access surgical OR keeps in mind four basic ideas: create the room design, utilize the equipment, integrate the information systems and put in place the team!

The Design

The Endosuite® operating room concept provides for patient contact equipment to be placed on easily movable booms, which are ceiling mounted. This arrangement allows for easy to clean rooms and thus faster turnovers. All patient contact equipment and monitors are placed on these booms while all other equipment is moved to the peripheral edge of the surgical suite, usually in a nurse's command and control center. This design puts control of all equipment at the fingertip of the circulating nurse and prevents the disruption and inconvenience of manipulating equipment on carts.

The easily movable booms make rearranging monitors simpler than moving carts, with the added advantage of having no wires to trip on, since the wiring is done through the boom structure. And since the wiring moves with the equipment, there is less chance of accidentally changed settings or unplugged wires, saving wear and tear on the equipment and staff.

Since the booms are easily moved to the periphery of the room, these rooms can easily be used for multi-specialty procedures, including non-minimally invasive procedures. The new operating suite is shown in Figure 4.1.

The Equipment

In addition to having the correct room design for a given facility, it is critical that the equipment selected provides the highest level of video quality and incorporates the latest developments in command and control systems. Special attention should be given to three areas of development in the OR of the Future: Visualization equipment, Computer Robotics, and Voice Activation of the OR. In our operating rooms, we utilize the Stryker Endoscopy 3-Chip video equipment, along with the Stryker/ Computer Motion HERMES Voice activation Command and Control surgical robot, the Computer Motion AESOP® Camera Positioning and Placement robot as well as the Stryker Communications Interactive Campus Telecommunications System.

Visualization

No other device is as critical to the success of a laparoscopic procedure as the video camera. Without proper visualization, there can be no accurate identification and treatment. Generally, two types of cameras are available for endoscopic surgery—1-chip or 3-chip video cameras. These cameras use solid-state, light-sensitive receptors called CCDs (charge-coupled devices, or "chips") that are able to detect brightness differences at different points throughout an image.

A 3-CCD camera (Fig. 4.2) uses a prism block that separates the full color image into its red, blue, and green components. Reflection off of a color-selecting mirror or passage through the mirror for one of the colors, in the prism directs each of these three colors to one of the three dedicated CCDs. White is represented by the presence of all three colors, while black is represented by the absence of all three colors. These 3-CCD cameras provide the greatest resolution and light sensitivity; however, they are also the most expensive.

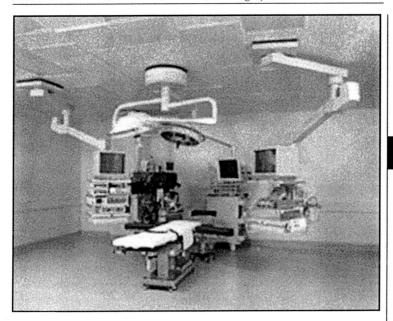

Fig. 4.1. The new Operating Room suites at Medical College of Virginia incorporate the Endosuite® design for Laparoscopic and Lap-Bariatric procedures.

The 1-CCD cameras use a single charge-coupled device with an overlay of millions of colored filters. Electronics within the camera or camera control unit are able to determine what filter the light hitting a specific point in the CCD is passing through. It is therefore possible to produce a smaller and less expensive camera than a 3-CCD camera; however, resolution and light sensitivity are both compromised.

The trend in surgery has definitely been toward smaller scope diameters, particularly a migration from 10 mm to 5 mm rigid scopes. As a result, utilizing a camera that can perform under the lower light parameters set forth under a 5 mm scope is essential. Moreover, the light source and its intensity must be factored in. When performing a laparoscopic procedure with 5 mm ports, it is often preferred to use a Xenon light source in order to maximize light throughput and guarantee pinnacle resolution.

Voice Activation and Computer Robotics in the Operating Room

There are some inherent shortcomings in the way operating room equipment has been traditionally accessed. The advent of minimally invasive procedures exacerbated these problems and underscored the need to improve upon the surgeon's avenues for integration. Because most of the MIS components reside outside the sterile field, the point person for critical controls became the circulating nurse. This led to increased frustration for the surgeon as well as the nurse. Often times, the

4

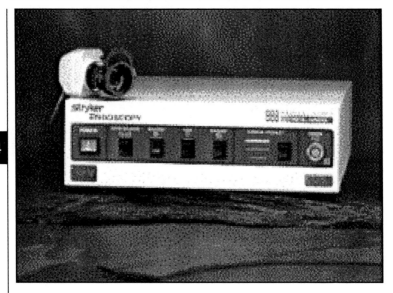

Fig. 4.2. Three chip digital video camera.

circulator would be out of the room at the precise moment that an adjustment, such as in the level of the insufflator's CO_2, had to be made. Surgeons grew frustrated at the subsequent delays, as well as an inability to take their own steps to change things. Additionally, nurses grew weary of such responsibilities. These constant interactions with the video tower pulled nurses away from patient-related tasks and from necessary clerical and operational work. The coup-de-grâce became the nurse's frustration with handling the exponentially higher complexity of video equipment and the invariable troubleshooting that ensued. The answer was certainly pointing toward improving the surgeon's access to these critical devices: voice activation.

Voice activation had been in the works since the late 1960s. Companies such as IBM began to look into developing software programs that would carry out human commands. The Holy Grail was, and continues to be, a simple, safe, and universally acceptable voice recognition system that flawlessly carries out the verbal requests of the user. However, the curve on this development proved steep. The ability to recognize a wide array of speech patterns was a technological hurdle that only today is showing true signs of promise. The good news is that voice recognition is here to stay and has begun to permeate many facets of the life of the everyday consumer. Voice control technology in automobiles, phone systems, and home environmental controls are but a few examples of this.

In 1998 the first FDA approved system for voice activation was introduced into the operating room. That system, known as HERMES, was the result of a co-development project between Stryker Endoscopy and Computer Motion. Designed to provide the surgeon with direct access and control of surgical devices, HERMES is operated via a hand-held pendant and/or surgeon voice commands.

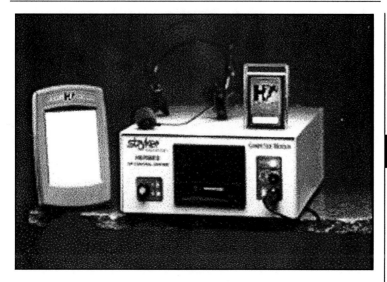

Fig. 4.3. The HERMES™ Command and Control System from Stryker Endoscopy.

HERMES™

The HERMES™ System (Fig. 4.3) gives direct control of surgical devices to the surgeon and provides the OR team with critical information. Surgeons have immediate access to "intelligent" medical devices using simple verbal commands or a hand-held touch-screen pendant. HERMES™ can be used in a broad array of minimally invasive surgical procedures.

To operate the HERMES™ device, the surgeon must take approximately 20 minutes to put his or her voice patterns on a PCMCIA (PC) card. This is accomplished by using a software program that walks the surgeon through a series of commands and captures segments of sounds called phonemes. These sound 'bits' are comprised of the pitch and inflection of how each syllable is formed by the user. When finished, the surgeon places the card inside the HERMES™ controller, which alerts him to the status of the system.

The operation of HERMES™ is relatively easy, with a learning curve of approximately 2-3 cases.

AESOP®

AESOP® (Fig. 4.4) is a voice-activated device, manufactured by Computer Motion, which is designed to hold a laparoscopic camera and scope. The surgeon can use several verbal commands to move the arm (and thus the camera) to keep up with the visual demands of the case. By utilizing a steady robotic arm, the surgeon is not adversely affected by unwanted movements and tremor of the image, often associated with human control of the camera. Moreover, there is less cause to clean the scope when it is controlled in a steadier and more deliberate fashion.

4

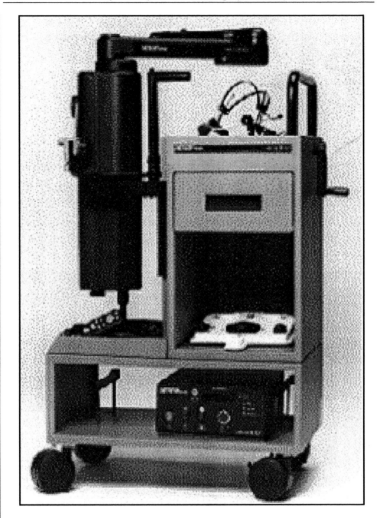

Fig. 4.4. AESOP® is a voice-activated device (Computer Motion, CA).

The Integration

The Stryker Communications Interactive Campus Telecommunications System (ICTS) is used our OR to provide voice, video and data communication to the OR from throughout the facility. A typical ICTS is used to connect the surgical OR to areas such as pathology, radiology, ER, conference rooms, surgeon offices, Wet/Dry Lab, etc. This provides two-way audio and video teleconferencing for the purpose of education, proctoring, consultation, viewing of X-rays or PACS images, viewing of gross specimen slides, for ER consultation, and for producing live educational courses

4

Fig. 4.5. Switchpoint III, a master audio/video and data router for the OR, provides a simple touchpanel interface for all equipment and information flowing into, out of, or routing in the Operating Room.

into our various conference rooms and auditoriums. These systems can also be used for connecting globally to anywhere around the world.

The ICTS interface in the OR is through the Switchpoint III (Fig. 4.5) which gives the circulating nurse or surgeon a simple touchpanel interface to control all equipment and information flow into or out of the room. The Stryker telecommunications system allows all departments to exchange video or data with personnel in the OR immediately, which provides both time and costs savings.

The Team

Much goes into the issue of time efficiencies within the OR. Excessive procedure, set-up and turnover time all contribute to a less than ideal productivity performance. As managed care continues to evolve, less efficient health care providers will be left behind. The right equipment and the right room design provide the basic foundation for a more productive OR. Without the right team, however, time efficiencies cannot be optimized. Turnover time and the flow of the operation are critically dependent on having a team orientation to minimally invasive procedures.

More hours probably have gone into studying turnover time than any other performance improvement activity in surgery. It is clear from these studies that a specially designated team of circulators and scrub techs for MIS procedures will save a significant amount of time as compared to a random group of circulators and scrub techs. These timesavings can be channeled into a larger volume of procedures, and a more relaxed, patient care driven environment.

Selected Readings

1.   Clemons, Bette J. The first modern operating room in America, AORN J, 01/2000.
2.   Patterson, Pat, Turnover time: Is all the study worth the effort?, OR Manager, 1999; 5.
3.   Fernsebner, Billie, RN, MSN, CNOR, Building a staffing plan based on OR's needs, OR Manager, 1996.

4

Effects of Increased Intra-Abdominal Pressure on Laparoscopic Surgery in Severe Obesity

Harvey J. Sugerman

Laparoscopic surgery has become very popular for the treatment of severe obesity. Obesity can be distributed in either an android fashion, primarily within the abdominal area or centrally as seen primarily in male patients, or in a gynoid manner, in the hips and buttocks, peripherally as seen primarily in female patients. Many of our severely obese female patients have both peripheral and central obesity. We have found that central obesity is associated with a significant increase in intra-abdominal pressure and this pressure is as high or higher than the pressure seen in patients with an "acute abdominal compartment syndrome" (Fig. 5.1). Data support the finding that this increase in intra-abdominal pressure is associated with a number of obesity related co-morbidity problems leading to the development of a "chronic abdominal compartment syndrome". These co-morbidities include obesity hypoventilation syndrome with its high cardiac filling pressures, gastroesophageal reflux disease, venous stasis disease, pseudotumor cerebri, an increased risk of incisional hernia and it is probably the cause of systemic hypertension and the nephrotic syndrome of obesity.

With regards to the specific problems of laparoscopic surgery associated with increased intra-abdominal pressure, there are several issues that need to be discussed. As the abdominal pressure is already elevated, especially in male patients, it may require a greater insufflation pressure than that used in thin patients in order to obtain an adequate pneumoperitoneum for enough visibility to perform the operation.

Animal studies have shown that acutely increased intra-abdominal pressure may lead to a decreased venous return to the heart with a decreased cardiac output primarily due to pressure on the inferior vena cava (Fig. 5.2). Further increases in intra-abdominal pressure can raise intra-thoracic pressure (Fig. 5.3), which will further compromise cardiac function and could cause severe hypotension. Thus, in patients with coronary artery atherosclerosis or carotid stenosis an acutely decreased venous return could lead to cardiovascular collapse, heart failure, myocardial infarction or stroke. Therefore, the surgeon and anesthesiologist need to be very observant of the patient's vital signs during abdominal insufflation. Intermittent pneumatic venous compression boots have been shown to increase venous return and counteract the effects on the lower body venous system. Should the patient become hemodynamically unstable, the pneumoperitoneum should be reduced, the patient given

Laparoscopic Bariatric Surgery: Techniques and Outcomes, edited by Eric J. DeMaria, Rifat Latifi and Harvey J. Sugerman. ©2002 Landes Bioscience.

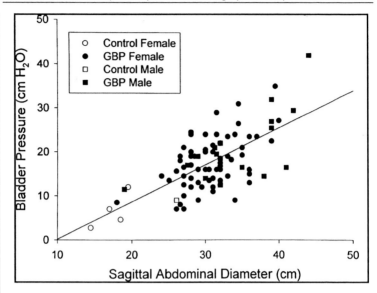

Fig. 5.1. Correlation between urinary bladder pressure and sagittal abdominal diameter in 84 morbidly obese patients and 5 "control" non-obese patients with ulcerative colitis, r = 0.67, p < 0.0001. Reprinted with permission from Sugerman HJ, Windsor ACJ, Bessos MK, Wolfe L. Abdominal pressure, sagittal abdominal diameter and obesity co-morbidity. J Int Med 1997; 241:71-9.

additional intravascular volume and the pneumoperitoneum re-established. If the patient again becomes hypotensive the patient should be converted to an open procedure. Pulmonary artery monitoring with measurement of mixed venous oxygen saturation using a co-oximeter catheter and cardiac output determinations, echocardiography or transesophageal Doppler sonography may be of benefit. Increased intra-thoracic pressure can decrease venous return from the brain (Fig. 5.4), producing an acutely increased intracerebral pressure (ICP). This is rarely a clinical problem; however, patients with a space-taking lesion in the brain could develop a further increase in ICP as noted in a case report of a patient with a cerebral neuroma undergoing laparoscopic cholecystectomy.

The decreased venous return from the inferior vena cava and lower extremities associated with an increased intra-abdominal pressure will put the patient at an increased risk of lower extremity venous thrombosis and pulmonary embolism. This might be further aggravated by a prolonged laparoscopic procedure. There is a concern that this risk may be increased in morbidly obese patients undergoing laparoscopic gastric bypass surgery. The increased intra-abdominal pressure in centrally obese patients presumably increases their risk of venous stasis disease prior to pneumoperitoneum. These patients require efforts to reduce the risk of venous thombosis and thomboembolism. Intermittent thigh-length venous compression stockings have been shown to increase venous return from the lower extremities in patients undergoing laparoscopy. Preoperative heparinization should also

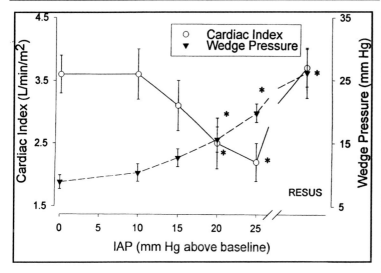

Fig. 5.2. Progressive increase in pulmonary artery wedge pressure and decrease in cardiac index with increasing intra-abdominal pressure associated with the intra-abdominal instillation of isosmotic polyethylene glycol in an acute porcine model. In order to return the cardiac index to baseline an estimated 1/3 of the animal's intravascular volume of lactated ringer's solution intravenous infusion was required which caused the wedge pressure to rise even further. *, $p < 0.05$ compared to baseline or pre-resuscitation values. Reprinted with permission from Ridings PC, Bloomfield GL, Blocher CR, Sugerman HJ. Cardiopulmonary effects of raised intra-abdominal pressure before and after intravascular volume expansion. J Trauma 1995; 39:1071-5.

be considered. This is probably even more important in patients undergoing laparoscopic Gastric Bypass.

Increased intra-abdominal pressure will push the diaphragm superiorly and compress the lungs. This increased intra-thoracic pressure is probably the cause for obesity hypoventilation syndrome and the increased cardiac filling pressures noted in patients with this complication of obesity. A prolonged laparoscopic procedure in the morbidly obese patient could lead to marked compressive atelectasis, which would be associated with postoperative pulmonary compromise and an increased risk of pneumonia. However, a recent randomized, prospective trial noted slightly improved pulmonary function following laparoscopic as compared to open gastric bypass. This is consistent with earlier data showing improved postoperative pulmonary function with a laparoscopic as compared to an open cholecystectomy. Nevertheless, an organized, expeditious approach to laparoscopic gastric bypass without prolonged operating times should be the goal of laparoscopic obesity surgery, supporting efforts to minimize the "learning curve" of this technically challenging procedure through intensive efforts through proctoring with experienced surgeons who have themselves advanced well beyond the "learning curve."

5

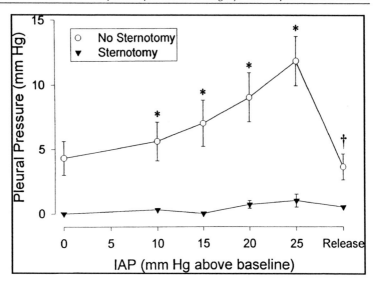

Fig. 5.3. Progressive increase in pleural pressure with increasing intra-abdominal pressure associated with the intra-abdominal instillation of isosmotic polyethylene glycol in an acute porcine model and prevention of this increase in animals who had undergone a median sternotomy and pleuropericardiotomy. Reprinted with permission from Bloomfield GL, Ridings PC, Sugerman HJ, Blocher CR. Increased pleural pressure mediates the effects of elevated intra-abdominal pressure upon the central nervous and cardiovascular systems. Crit. Care. Med.1997; 25:496-503.

There are data in animal models showing that increased intra-abdominal pressure can lead to a decreased splanchnic as well as hepatic perfusion associated with a decreased portal venous flow. There are case reports of intestinal infarction following laparoscopic surgery; however, there have not been any reports to date of this complication following laparoscopic gastric bypass. Superior mesenteric vein, portal vein thrombosis or hepatic vein thrombosis could be rare complications of acutely increased intra-abdominal pressure. These are other reasons to minimize operating times with laparoscopic surgery.

Increased intra-abdominal pressure will increase the pressure on the renal veins and activate the renin-angiotenin-aldosterone system (RAAS) and may be the cause for systemic hypertension and proteinuria in obesity. Theoretically, this could cause a marked increase in blood pressure during laparoscopic surgery; however, this does not seem to be a clinical problem. A prolonged increase in renal venous pressure could lead to renal vein thrombosis and its complication of nephrotic syndrome and hypertension.

In summary there are a number of potential complications associated with a prolonged and high level of intra-abdominal pressure, which can affect virtually every organ of the body. Thus, the goal should be to perform laparoscopic surgery with as low an intra-abdominal pressure as possible and as expeditiously as reasonable without jeopardizing patient safety by excessive speed.

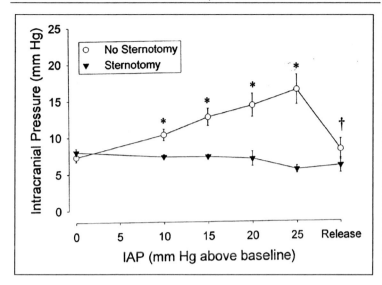

Fig. 5.4. Progressive increase in directly measured intracranial pressure with increasing intra-abdominal pressure associated with the intra-abdominal instillation of isosmotic polyethylene glycol in an acute porcine model and prevention of this increase in animals who had undergone a median sternotomy and pleuropericardiotomy. Reprinted with permission from Bloomfield GL, Ridings PC, Sugerman HJ, Blocher CR (1997): Increased pleural pressure mediates the effects of elevated intra-abdominal pressure upon the central nervous and cardiovascular systems. Crit Care Med 1997; 25:496-503.

Selected Readings

1. Sugerman HJ, Windsor ACJ, Bessos MK, et al. Abdominal pressure, sagittal abdominal diameter and obesity co-morbidity. J Int Med 1997; 241:71-9.
2. Ridings PC, Bloomfield GL, Blocher CR, et al. Cardiopulmonary effects of raised intra-abdominal pressure before and after intravascular volume expansion. J Trauma 1995; 39:1071-5.
3. Barnes GE, Laine GA, Giam PY, et al. Cardiovascular response to elevations of hydrostatic intraabdominal pressure. Am J Physiol 1095; 248:R208-13.
4. Goodale RL, Beebe DS, McNevin MP, et al. Hemodynamic, respiratory and metabolic effects of laparoscopic cholecystectomy. Am J Surg 1993; 166:533-7.
5. PunnonenR, Viinaki O. Vasopressin release during laparoscopy: Role of increased intra-abdominal pressure. Lancet 1982; 8264:175-6.
6. Kelman GR, Swapp GH, Smith I, et al. Cardiac output and arterial blood-gas tension during laparoscopy. Br J Anaesth 1972; 44:1155-61.
7. Brantly JC, Riley PM. Cardiovascular collapse during laparoscopy: A report of two cases. Am J Obstet Gynecol 1998; 159:735-7.
8. Westerband A, Van De Water JM, Amzallag, M, et al. Cardiovascular changes during laparoscopic cholecystectomy. Surg Gynecol Obstet 1992; 175:535-8.

5

9. Alishahi S, Francis N, Crofts S, Duncan L, et al. Central and peripheral adverse hemodynamic changes during laparoscopic surgery and their reversal with a novel intermittent sequential pneumatic compression device. Ann Surg 2001; 233:176-81.

10. Loder WA, Minnich M, Brotman S. Hemodynamic effects of laparoscopic cholecystectomy. Am Surg 1994; 60:322-5.

11. Elliot S, Savil P, Eckersall S. Cardiovascular changes during LC: a study using transesophageal Doppler monitoring. Eur J Anesthesiol 1998; 15:50-5.

12. Portera CA, Compton RP, Walters DN, Browder IW. Benefits of pulmonary artery catheter and transesophageal echocardiographic monitoring in laparoscopic cholecystectomy patients with cardiac disease. Am J Surg 1995; 169:202-6.

13. Sugerman HJ, DeMaria EJ, Felton WL III, et al. Increased intra-abdominal pressure and cardiac filling pressures in obesity associated pseudotumor cerebri. Neurology 1997; 49, 507-11.

14. Bloomfield GL, Ridings PC, Sugerman HJ, et al. Increased pleural pressure mediates the effects of elevated intra-abdominal pressure upon the central nervous and cardiovascular systems. Crit Care Med 1997; 25:496-503.

15. Irgau I, Koyfman Y, Tikellis JI. Elective intraoperative intracranial monitoring during laparoscopic cholecystectomy. Arch Surg 1995; 130:1101-3.

16. Beebe DS, McNevin MP, Crain JM, et al. Evidence of venous stasis after abdominal insufflation for laparoscopic cholecystectomy. Surg Gynec Obstet 1993; 176:443-7.

17. Mayol J, Vincent-Hamelin E, Sarmiento JM, et al. Pulmonary embolism following laparoscopic cholecystectomy: Report of two cases and review of the literature. Surg Endosc 1994; 8:214-7.

18. Ido K, Suzuki T, Taniguchi Y, et al. Femoral vein stasis during laparoscopic cholecystectomy: Effects of graded elastic compression leg bandages in preventing thrombus formation. Gastrointest Endosc 1995; 42:151-5.

19. Lindberg F, Bergqvist D, Rasmussen I. Incidence of thromboembolic complications after laparoscopic cholecystectomy: Review of the literature. Surg Laparosc Endosc 1997; 7:324-31.

20. Nguyen NT, Luketich JD, Friedman DM, et al. Pulmonary embolism following laparoscopic antireflux surgery: A case report and review of the literature. JSLS 1999; 3:149-53.

21. Sugerman HJ, Sugerman EL, Wolfe L, et al. Risks/benefits of gastric bypass in morbidly obese patients with severe venous stasis disease. Ann Surg 2001; 234:41-6.

22. Millard JA, Hill BB, Cook PS, et al. Intermittent sequential pneumatic compression in prevention of venous stasis associated with pneumoperitoneum during laparoscopic cholecystectomy. Arch Surg 1993; 128:914-9.

23. Schwenk W, Böhm B, Fügener A, et al. Intermittent sequential pneumatic compression (ISC) of the lower extremities prevents venous stasis during laparoscopic cholecystectomy. Surg Endosc 1998; 12:7-11.

24. Sugerman HJ, Fairman RP, Sood RK, et al. Long-term effects of gastric surgery for treating respiratory insufficiency of obesity. Am J Clin Nutr 1992; 55:597S-601S.

25. Sugerman HJ, Baron PL, Fairman RP, et al. Hemodynamic dysfunction in obesity hypoventilation syndrome and the effects of treatment with surgically induced weight loss. Ann Surg 1988; 207:604-13.

26. Ridings PC, Bloomfield GL, Blocher CR, et al. Cardiopulmonary effects of raised intra-abdominal pressure before and after intravascular volume expansion. J. Trauma 1995; 39:1071-5.

27. Nguyen NT, Lee SL, Goldman C, Ho HS, et al. Comparison of pulmonary function and postoperative pain after laparoscopic and open gastric bypass: A randomized trial. J Am Coll Surg, 2001; 192:469-7.

28. Schauer PR, Luna J, Ghiatas AA, et al. Pulmonary function after laparoscopic cholecystectomy. Surgery 1993; 114:389-99.
29. Schauer PR, Ikramuddin S, Ramanathan R, Gourash W, Panzak G. Outcomes after laparoscopic Roux-en-Y gastric bypass for morbid obesity. Ann Surg, 2000; 232:515-2.
30. Diebel LN, Wilson RE, Dulchavsky SA, et al. Effect of increased intra-abdominal pressure on hepatic arterial, portal venous, and hepatic microcirculatory blood flow. J Trauma. 1992; 33;279-84.
31. Jakimowicz J, Stultiëns G, Smulders F. Laparoscopic insufflation of abdomen reduces portal venous flow. Surg Endosc 1998; 12:129-32.
32. Junghans T, Böhm B, Gründel K, et al. Does pneumoperitoneum with different gases, body positions, and intraperitoneal pressure influence renal and hepatic blood flow? Surgery 1997; 121:206-10.
33. Pelosi MA III, Pelosi MA. Laparoscopic-associated intestinal infarction: A new syndrome? Am J Obstet Gynecol 1999; 180:773-4.
34. Millikan KW, Szezerba SM, Dominguez JM, et al. Superior mesenteric and portal vein thrombosis following laparoscopic-assisted right hemicolectomy. Report of a case. Dis Colon Rectum 1996; 39:1171-5.
35. Bloomfield GL, Blocher CR, Sugerman HJ. Elevated intra-abdominal pressure increases plasma renin activity and aldosterone levels. J Trauma 1997; 42:997-1004.
36. Doty JM, Saggi BH, Sugerman HJ, et al. The effect of increased renal venous pressure on renal function. J Trauma 1999; 47:1000-4.
37. Bloomfield GL, Sugerman HJ, Blocher CR, et al. Chronically increased intra-abdominal pressure produces systemic hypertension in dogs. Int J Obes Relat Metabol Disord 2000; 24:819-24.
38. Lopez Cubillana P, Asensio Egea LJ, Rigabert Montiel M, et al. Renal vein thrombosis caused by laparoscopic cholecystectomy. Scand J Urol Nephrol 1998; 32:56-7.

CHAPTER 6

Prophylaxis of Venous Thromboembolism in Morbidly Obese Patients

Giselle G. Hamad

Introduction

The prevalence of obesity in the United States has achieved epidemic proportions. About 55% of Americans are overweight or obese and over one-third of the adult population is obese. The medical conditions associated with morbid obesity involve multiple organ systems and include type 2 diabetes mellitus, hypertension, hypertrophic cardiomyopathy, hyperlipidemia, obstructive sleep apnea, cholelithiasis, osteoarthritis, deep venous thrombosis (DVT), and pulmonary embolism (PE). There is a substantial increase in mortality in this population; morbidly obese young male adults have a 12-fold increase in mortality.

Deep venous thrombosis and pulmonary embolism are serious, potentially life-threatening disorders which contribute to significant morbidity and mortality. In the United States, there are between 1.5 and 2.5 million cases of DVT per year. There is an estimated incidence of 600,000 cases per year of PE in the United States with up to 200,000 mortalities. PE accounts for 15% of postoperative deaths. These statistics probably underestimate the true incidence of DVT and PE, which may elude diagnosis.

The association between DVT and PE was described in 1856 by Virchow, who ascribed the development of DVT to the triad of stasis, endothelial damage, and hypercoagulability state. A study of risk factors for venous thromboembolism in hospitalized patients demonstrated an association with age over 40 years (59%), obesity (28%), and major surgery (23%). The increased risk in the morbidly obese is attributable to a sedentary lifestyle because of difficulty ambulating and the substantial amount of weight resting on the inferior vena cava. Additional risk factors include prior history of DVT or PE, immobility, pregnancy, oral contraceptive use, smoking, hypercoagulable states, malignancy, anesthesia, trauma, and orthopedic surgery (Table 6.1). Although they are usually diagnosed as postoperative complications, DVT and PE may also occur in non-hospitalized patients.

Patients with morbid obesity who undergo surgical procedures under general anesthesia are at substantial risk for development of DVT and PE. In a study from Sweden in 1997, among 328 patients undergoing bariatric surgery, the incidence of venous thromboembolism was 2.4% despite prophylaxis with dextran or subcutaneous unfractionated heparin. Whether laparoscopic surgery directly increases the risk of DVT and PE is unclear. DVT and PE have been reported following

Laparoscopic Bariatric Surgery: Techniques and Outcomes, edited by Eric J. DeMaria, Rifat Latifi and Harvey J. Sugerman. ©2002 Landes Bioscience.

Table 6.1. Risk factors for deep venous thrombosis and pulmonary embolism

Age over 40 years
Obesity
General anesthesia
Prior history of DVT or PE
Immobility
Pregnancy
Oral contraceptive use
Smoking
Hypercoagulability
Malignancy
Trauma
Hip and knee replacement

---

laparoscopic cholecystectomy, and venous stasis is an important contributor. Millard and colleagues demonstrated that a 42% reduction in femoral venous flow occurred during laparoscopic cholecystectomy concomitant with pneumoperitoneum and reverse Trendelenburg; this effect was reversed with use of sequential pneumatic compression devices. Patients undergoing laparoscopic bariatric surgery are subjected to prolonged reverse Trendelenburg positioning and pneumoperitoneum intraoperatively, both of which increase venous stasis and further increase the risk of DVT and PE. The reverse Trendelenburg position is also used postoperatively for optimal pulmonary function. A number of studies demonstrating a hypercoagulable state associated with laparoscopy in both animals and humans have surfaced, underscoring the importance of effective prophylaxis and early postoperative ambulation.

Pulmonary Embolism and Obesity

Two studies of risk factors of PE in the general population have demonstrated an association with obesity among females. The Framingham study in 1983 identified long-term risk factors for major pulmonary embolism in 46 autopsy-confirmed pulmonary embolism cases from the Framingham Heart Study. Univariate analysis demonstrated that only Metropolitan relative weight was significantly and independently associated with PE, and only among females. In a prospective study of risk factors for pulmonary embolism in women, the risk of primary PE was associated with obesity, cigarette smoking, and hypertension in multivariate analysis.

A study of 855 men from a random population sample from Sweden in 1999 identified smoking and abdominal obesity as independent risk factors for the development of venous thromboembolism.

In a Mayo clinic study of 36 autopsy subjects who died of PE without previously recognized clinical or environmental risk factors, 67% were morbidly obese. Compared to age- and gender-matched controls, body mass indices were significantly higher in the subjects who died of PE. The data suggest that morbid obesity is an independent risk factor for pulmonary embolism.

Coagulation Abnormalities in Obesity

Obesity is associated with a number of derangements in coagulation (Table 6.2). Plasma concentrations of fibrinogen, von Willebrand factor antigen, tissue-type plasminogen activator (t-PA) antigen, and factor VII contribute to a hypercoagulable state. Platelet aggregation appears to be augmented, and leptin has been implicated as a promoter of platelet aggregation. Upregulation of plasminogen activator inhibitor-1 (PAI-1), an inhibitor of fibrinolysis, is frequently seen in obesity; adipose tissue has been shown to be a source of PAI-1 secretion, which is stimulated by transforming growth factor beta (TGF-β) Plasma levels of PAI-1 antigen and von Willebrand factor antigen have been demonstrated to correlate with central obesity in women.

Weight reduction leads to normalization of several coagulation parameters. A study by Batist and colleagues in 1983 of 23 morbidly obese adults demonstrated a deficiency of antithrombin III compared to normal weight controls; levels of this endogenous anticoagulant normalized with weight reduction. Thirty-six obese patients who lost an average of 13.6 kg by dietary measures had a significant reduction in factor VII coagulant activity and plasminogen activator inhibitor-1 (PAI-1) antigen. Folsom et al demonstrated significant reductions in t-PA antigen, PAI-1 antigen, and factor VII correlating with amount of weight lost. In a study by Primrose and colleagues, surgery for morbid obesity resulting in a mean weight loss of 64 kg at 12 months was accompanied by significant decreases in fibrinogen, factor VII, and PAI-1 activity. Therefore, the dietary and surgical treatment of morbid obesity may reduce the mortality from cardiovascular and thromboembolic disease.

Perioperative Prophylaxis

Venous thromboembolism is a major cause of mortality among morbidly obese patients undergoing bariatric surgery. The majority of bariatric surgeons use prophylaxis routinely. Despite strict adherence to perioperative prophylactic measures, complications of venous thrombosis and pulmonary embolism continue to plague these high risk patients. Currently, there is no consensus with regards to the best method for prophylaxis against venous thromboembolism in morbidly obese patients undergoing bariatric surgery. A recent survey of members of the American Society for Bariatric Surgery demonstrated a self-reported incidence of 2.63% for DVT and 0.95% for PE. Greater than 95% of surgeons routinely used prophylaxis for venous thromboembolism; the preferred methods included unfractionated heparin, intermittent pneumatic compression stockings, and low molecular weight heparins. Forty-eight percent of surgeons reported at least one death from PE in their bariatric surgical practice.

Early postoperative mobilization, elastic compression stockings, and intermittent pneumatic compression devices are simple and noninvasive interventions. The stockings should be placed upon the induction of anesthesia and are worn until the patient ambulates. Proper fit is mandatory. These devices enhance venous return, reduce venous stasis, and stimulate fibrinolysis. In reality, however, they are cumbersome and uncomfortable for some patients and therefore may be underutilized in the postoperative period.

Subcutaneous unfractionated heparin has been widely advocated for postoperative prophylaxis against venous thromboembolism. However, the recommendation for dosing in nonobese adults (5000 units every 8 to 12 hours) may be inadequate in obese adults, who may be undertreated with this regimen. Complications of prophylaxis with unfractionated heparin include heparin-induced thrombocytopenia and hemorrhage.

The efficacy of low molecular weight heparins (LWMHs) in the prevention of venous thromboembolism is well-established. LWMHs have a more predictable effect on coagulation, better bioavailability, and a longer elimination half-life than

Table 6.2. Coagulation abnormalities in obesity

Elevated fibrinogen
Elevated von Willebrand factor antigen
Elevated tissue-type plasminogen activator (t-PA) antigen
Elevated factor VII
Enhanced platelet aggregation
Elevated plasminogen activator inhibitor-1 (PAI-1)
Deficiency in antithrombin-III

unfractionated heparin. LMWHs have been shown to be superior to unfractionated heparin or warfarin in the prophylaxis against venous thromboembolism in patients undergoing orthopedic surgical procedures, with fewer bleeding complications. LMWH was more effective than unfractionated heparin in preventing venous thromboembolism in a prospective, randomized clinical trial of patients with major trauma. Among general surgery patients, LMWH has been shown to be as effective and safe as unfractionated heparin in the prevention of postoperative deep venous thrombosis after general surgical procedures in a randomized multicenter study. The dose of enoxaparin recommended for patients undergoing abdominal surgery who are at risk for thromboembolic complications is 40 mg SC qd.

Among morbidly obese patients, it is unclear whether non-weight based regimens for dosing of LMWH may be safely applied. Underdosing these patients with LMWH is likely to increase their risk of thromboembolic complications. LMWHs induce a weaker prolongation of the activated partial thromboplastin time (APTT) than UH; therefore APTT is not useful for monitoring LMWH therapy. The most commonly used tests for monitoring therapy with LMWH measure the anti-factor Xa plasma activity, but this parameter is not routinely monitored for thromboprophylaxis.

A number of authors have demonstrated a correlation between anti-Xa activity and body weight with LMWH administration for thromboprophylaxis. In a prospective study of 547 patients undergoing gynecologic surgery, anti-Xa activity following enoxaparin administration was found to correlate with body weight. A similar correlation was demonstrated in a study of thromboprophylaxis with reviparin in 42 pregnant patients, suggesting that doses of LMWH may require adjustment according to body weight. Vitoux and colleagues studied 51 medical inpatients receiving fraxiparine, a LMWH, for thromboprophylaxis. Mean body weight was 63.8 kg (range 35-95). The response to a constant dose of 0.3 ml was followed with anti-factor Xa levels obtained 3 hours after injection. There was a significant negative correlation (P<0.001) between body weight and anti-Xa values; the authors suggested that the safety of prophylaxis with LMWH might be improved by adjusting dosage according to body weight.

For a select group of patients who are considered to be at extremely high risk for thromboembolism, preoperative prophylactic insertion of a vena caval filter has been advocated by some bariatric surgeons. However, radiographic visualization for filter placement may be limited in obese patients. Sugerman advocates prophylactic insertion of a vena caval filter in patients with respiratory insufficiency of obesity and a mean pulmonary artery pressure of 40 mm Hg or higher.

Diagnosis

DVT is often a challenge to diagnose because physical examination is often unreliable. Clinical findings associated with DVT include fever, calf pain and tenderness, lower extremity edema, erythema, warmth, pain induced by calf dorsiflexion (Homan's sign), and a palpable venous cord. Fewer than 30% of patients diagnosed with PE present with clinical signs of lower extremity DVT. Clinical manifestations of PE include dyspnea, chest pain, tachycardia, tachypnea, hypotension, fever and hemoptysis. If PE is recognized and treated promptly, mortality is 2.5%; if it is unrecognized and therefore untreated, the mortality rises to 30%.

Postoperative gastric bypass surgery patients with tachycardia, chest pain, fever, and respiratory distress should also be suspected of having a leak from the gastrojejunal anastomosis, a complication which has life-threatening potential and mandates immediate attention. Therefore, whether an evaluation for either pulmonary embolism or anastomotic leak will be instituted first should be dictated by the clinical scenario. If the patient's signs and symptoms strongly suggest PE, a chest CT or V/Q scan should be obtained promptly. However, if an anastomotic leak is suspected, either an UGI or in some cases, immediate exploration must be undertaken.

Ventilation-perfusion (V/Q) scanning is commonly used as a first-line imaging modality for the detection of PE. Its interpretation must be considered in conjunction with the clinician's suspicion of the diagnosis of PE, but 20% of PEs may be missed despite this combination. Furthermore, up to 70% of scans are reported as indeterminate. Therefore, the goal of V/Q scanning is to identify patients at high risk for thromboembolic events who will require anticoagulation. Patients with intermediate probability V/Q scans often require further confirmatory testing to establish the diagnosis of PE. The Prospective Investigation of Pulmonary Embolism Diagnosis (PIOPED) study in 1990 found that 88% of patients with high-probability V/Q scans had PE, as did 12% with low-probability scans.

Spiral computerized tomography (CT) scans of the chest has emerged as the first test of choice and has high sensitivity and specificity. A study from Blachere et al in 2000 compared CT with V/Q scanning for the diagnosis of PE and found significantly higher sensitivity, specificity, positive- and negative-predictive value for CT. Confirmation of the diagnosis of PE with CT may not be possible in morbidly obese patients because of the physical limitations of the imaging equipment, which may not accommodate patients with a large abdominal girth. CT of the chest is also useful in the evaluation of nonvascular abnormalities involving the pleura and pulmonary parenchyma. This imaging modality involves intravenous injection of contrast dye and therefore should be avoided in patients with renal insufficiency or dye allergy.

Pulmonary angiography has been considered the gold standard for the diagnosis of PE. It is a costly, invasive study which involves intravenous contrast injection with the potential for nephrotoxicity.

Newer studies for the diagnosis of PE include the enzyme-linked immunosorbent assay (ELISA) for D-dimer and magnetic resonance angiography imaging. The rapid ELISA D-dimer assay is a noninvasive blood test which detects fibrin proteolysis and intravascular thrombus formation; low concentrations of D-dimer are thought to exclude PE or DVT. Because of its high negative predictive value, it is useful in excluding the diagnosis of PE. Magnetic resonance imaging shows promise as a noninvasive diagnostic study but will require further evaluation to determine its utility.

Therapy

Heparin prevents not only the extension of existing thrombus but also the formation of new thrombus. Treatment of PE and DVT with intravenous unfractionated heparin presents a formidable challenge in the morbidly obese because of a increased heparin dosage requirement and a lack of consensus on dosing guidelines. Recommendations for dosage of intravenous heparin have been based on actual body weight, ideal body weight, or blood volume. Complications of heparin therapy include heparin-induced thrombocytopenia and hemorrhage.

Acute DVT or PE may be managed with low molecular weight heparin, which is equal in efficacy as unfractionated heparin. However, standard dosing (1 mg/kg SQ BID) applied to morbidly obese patients may lead to massive overdoses and an increased risk of hemorrhage.

Treatment with oral warfarin may present a problem in postoperative gastric bypass patients whose oral intake has been drastically limited by their gastrointestinal neoanatomy. Patients may have an rapid escalation in their prothrombin time because of a dietary deficiency in vitamin K. Prothrombin times should be monitored frequently after treatment with warfarin has been initiated. Generally, treatment with warfarin is continued for 3 to 6 months.

Conclusion

Surgical patients with morbid obesity are at substantial risk of developing DVT and PE. Obesity has been shown to be an independent risk factor for the development of PE in several studies. A number of coagulation abnormalities in morbid obesity contribute to the elevated risk of thrombosis. Loss of excess weight leads to normalization of some of these coagulation parameters, thereby reducing the risk of thromboembolic disease. Prevention of thromboembolism in the perioperative period is critical, but there is no consensus as to the ideal method for prophylaxis in the morbidly obese surgical patient. Establishing the diagnosis of PE may be problematic in the morbidly obese. In the gastric bypass patient, the possibility of an anastomotic leak must be considered. Treatment of PE with heparin (unfractionated or low molecular weight heparin) and warfarin must be monitored closely.

Selected Readings

1. Value of the ventilation/perfusion scan in acute pulmonary embolism. Results of the prospective investigation of pulmonary embolism diagnosis (PIOPED). The PIOPED Investigators. JAMA 1990; 263:2753-9.
2. Blachere H, Latrabe V, Montaudon M, et al. Pulmonary embolism revealed on helical CT angiography: comparison with ventilation-perfusion radionuclide lung scanning. AJR Am J Roentgenol 2000; 174:1041-7
3. Ellison MJ, Sawyer WT, Mills, TC. Calculation of heparin dosage in a morbidly obese woman. Clin Pharm 1989; 8:65-8.
4. Eriksson S, Backman L, Ljungstrom K-G. The incidence of clinical postoperative thrombosis after gastric surgery for obesity during 16 years. Obesity Surgery 1997;7:332-335.

5. Marckmann P, Toubro S, Astrup A. Sustained improvement in blood lipids, co-agulation, and fibrinolysis after major weight loss in obese subjects. Eur J Clin Nutr 1998; 52:329-33.

6. Persson AV, Davis RJ, Villavicencio JL. Deep venous thrombosis and pulmonary embolism. Surg Clin N Am 1991; 71:1195-1209.

7. Sugerman HJ. Obesity. Care of the Surgical Patient. Wilmore DW, Brennan MF, Harken AH, Holcroft JW, Meakins JL, eds. Scientific American, New York, 1989.

8. Tapson VF. Pulmonary embolism—New diagnostic approaches. N Engl J Med 1997; 336:1449-51

9. Vitoux JF, Aiach M, Roncato M, et al. Should thromboprophylactic dosage of low molecular weight heparin be adapted to patient's weight? Thromb Haemostasis 1988; 59:120.

10. Wu EC, Barba CA. Current practices in the prophylaxis of venous thromboembolism in bariatric surgery. Obesity Surgery 2000; 10:7-14.

6

CHAPTER 7

Laparoscopic Adjustable Silicone Gastric Banding

Eric J. DeMaria and Rifat Latifi

Background

Gastric banding has been promoted as a treatment for obesity by many surgeons over recent decades. Advantages included technical ease of performing the procedure and no intestinal anastomosis with the added risks of anastomotic leak. The most popular banding procedure is the vertical-banded gastroplasty, however this involves creation of a gastric staple line with risks of staple line disruption. Gastric banding involves creating a gastric pouch by encircling the stomach with some type of material such as dacron, silastic, etc. in order to create a narrowed efferent tract. Possible complications of this procedure are erosion of the band material into the stomach lumen and intractable postoperative vomiting, if the patient does not follow dietary recommendations including eating slowly and careful chewing of food before swallowing.

Laparoscopic adjustable silicone gastric banding (LASGB) is a relatively new surgical procedure for the treatment of morbid obesity. The LASGB device (Fig. 7.1) clearly has advantages over other forms of gastric banding since there is a lower risk of eroding the gastric wall and the vomiting is not as prevalent. The device encircles the proximal stomach and is connected by tubing to a reservoir implanted and secured to the abdominal fascia in the patient's upper abdomen which can be accessed via a needle to inflate or deflate the band device. Thus, if vomiting develops as a result of band tightening in pursuit of weight loss, it may be relieved by band deflation should dietary counseling fail to alleviate the symptoms, avoiding reoperation. The LASGB procedure is clearly an easier laparoscopic procedure than laparoscopic gastric bypass.

Technique

Patient's Positioning and Trocar Placement

The laparoscopic adjustable band procedure is performed using a variety of techniques that share some common features.

Many surgeons place the patient in lithotomy position and stand between the patient's legs using trocars inserted in the left upper abdomen to perform the procedure. This is facilitated by placing a zero degree viewing laparoscope in a subxiphoid position to visualize the proximal stomach, to avoid the constant instrument

Laparoscopic Bariatric Surgery: Techniques and Outcomes, edited by Eric J. DeMaria, Rifat Latifi and Harvey J. Sugerman. ©2002 Landes Bioscience.

7

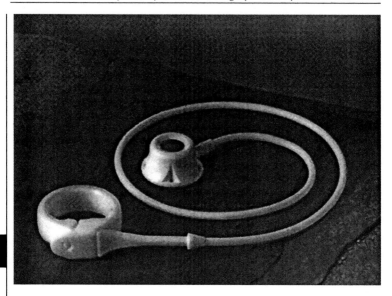

Fig. 7.1. The Lap-Band system (BioEnterics, Corp, Carpinteria, CA) ring for peri-gastric placement with demonstrated reservoir filled with saline to reduce stomal diameter.

'conflicts' created by a supra-umbilical placement adjacent to the surgeon's working port sites. While ergonomically correct, lithotomy is a difficult and awkward position entailing some risk to the morbidly obese patient.

Subsequent experience performing laparoscopic procedures in the morbidly obese has proven that it is possible to perform such procedures with the patient in a supine position with the surgeon working from the patient's right side with the assistant in the left side. For this we place the patient's right arm to his side with careful padding for protection to allow the surgeon flexibility in his/her positioning. In general, port insertion is similar to that of laparoscopic Nissen fundoplication. Retraction of the of the liver and exposure of the proximal stomach is achieved by Nathanson liver retractor, which is placed through a subxiphoid puncture, and which is anchored to the table using a rigid arm.

Steep reverse Trendelenburg positioning is required in the morbidly obese patient as a significant amount of intra-abdominal fat may obscure visualization of the proximal stomach, particularly omentum hanging from the greater curvature of the stomach. Occasionally we have placed a long suture through the superior-most 'tongue' of fat and brought it out through a left lateral trocar in order to facilitate this retraction. In order to safely tilt the OR table into such steep positions, a foot board must be placed and the legs and feet securely positioned and padded for protection to prevent the disastrous consequences of an obese patient sliding off the end of the table.

Dissection

The dissection is begun high on the lesser curvature of the stomach. The location of the dissection is critical in order to create an appropriately small proximal gastric pouch above the band. To aid in choosing the appropriate site for dissection, a balloon catheter (BioEnterics, Corp, Carpinteria, California) is advanced into the stomach, inflated with 25 ml of saline, and withdrawn until it becomes lodged at the gastroesophageal junction. The balloon is easily visualized laparoscopically within the gastric lumen. The site for dissection is at the balloon's equator with the goal of creating a 15 ml pouch. Initially, the peritoneum overlying the angle of His above the short gastric vessels is opened. Lesser curvature dissection proceeding behind the proximal stomach is the most difficult part of the procedure. The goal is to create a tunnel behind the stomach without entry into the free retrogastric space of the lesser sac. We believe that positioning of the band within this posterior tissue decreases the risk of posterior gastric herniation through the band, which results if the band is placed lower in the free space. For this reason we fix the band in position posteriorly with sutures in the event the lesser sac is entered during the dissection. The retrogastric tunnel should optimally be slightly larger in diameter than the band device itself. Dissection should follow alongside the shiny white tissue of the posterior stomach until the angle of His is reached laterally. A special retractor or reticulating grasper is then introduced into this retrogastric tunnel from the lesser curvature side of the stomach and extended so that the tip of the instrument is easily visualized laterally.

Placement of the Band

The band device itself is then placed into the peritoneal cavity through the 15 mm port with the tail of the band device introduced first attached to a grasper. The tubing of the band is then threaded through the port followed ultimately by inserting the inflatable band using a grasper attached to the plastic buckle. The band's tail is grasped internally and delivered to the reticulating instrument that is used to pull the band into position behind the stomach. Care must be taken to avoid tying a knot in the long tubing while the band is delivered into position.

Band closure is accomplished by inserting the band's tail into the buckle and pulling it through until the locking mechanism engages (Fig. 7.2). A closing tool facilitates this maneuver. The band must be correctly positioned at the moment of closure such that an appropriately small gastric pouch is created. A balloon catheter connected to a pressure sensor (Gastrostenometer, BioEnterics, CA) which registered the band closure tension facilitated this placement, but proved to be unnecessary with additional experience. The band is optimally closed against the inferior rim of the inflated 15 ml balloon.

Once the band is closed in position, three to four anterior gastro-gastric sutures are placed in the stomach proximal and distal to the band to secure it anteriorly and decrease the risk of anterior gastric herniation (Fig. 7.3). The band reservoir is left empty to decrease the early postoperative risk of vomiting which may increase the risk of early herniation. The device's tail is removed and the subcutaneous reservoir attached to the tubing, brought out through the left mid abdomen (15 mm port site). The reservoir port is anchored securely to the anterior abdominal fascia of the

7

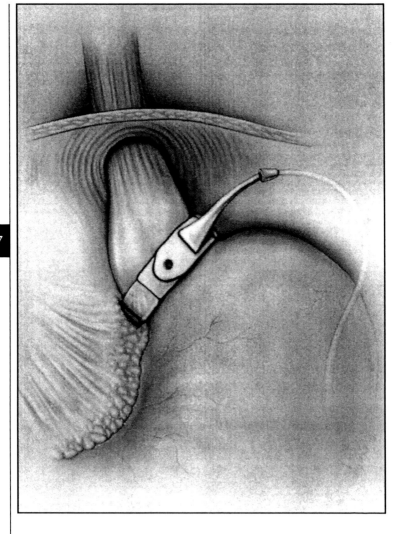

Fig. 7.2. The Lap-Band device positioned around the proximal stomach.

rectus sheath at this site using nonabsorbable sutures with open surgical technique. We close all large port site.

The skin wounds are irrigated, local anesthetic solution injected for postoperative analgesia, and closed. No intraperitoneal drains are left.

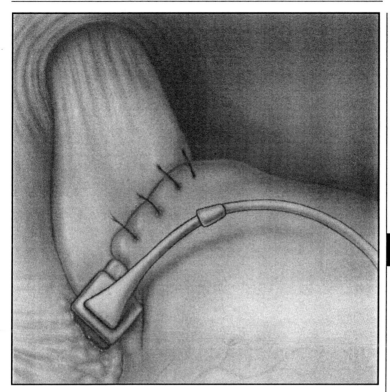

Fig. 7.3. The Lap-Band secured anteriorly with 3-4 gastro-gastric sutures to prevent gastric herniation through the band anteriorly.

Immediate Postoperative Management

Early postoperative ambulating is encouraged. A barium swallow is performed the morning after surgery to confirm correct band positioning and to rule out gastric injury. If this examination is normal, the patient is allowed to begin oral liquids and discharged from the hospital within 24 hours of the procedure. A pureed diet is usually tolerated within a few days of the surgery and is continued for a month. The patient should be cautioned that over-eating, nausea and vomiting might be dangerous.

Late Postoperative Care

Band adjustments should begin no less than one month following successful surgical placement. This is done to allow for formation of a pseudo-capsule around the perigastric band to decrease the risk of early postoperative band slippage or gastric herniation. The strategies for band adjustment varies. Narrowing the lumenal diameter of the band by injection of saline requires sterile technique and, sometimes fluoroscopic guidance. Up to a total of 4 ml of saline can be injected into the reservoir for maximal constriction before concerns arise about damage to the band from over-inflation. A logical adjustment strategy is to progressively narrow the band diameter until the patient begins a steady and sustained weight loss. One approach is to inject 1/2 ml of saline into the band at intervals of 2-4 weeks between injections while monitoring the patient's intake of both calories and food groups. Patients must be repeatedly told to avoid sugar and other sweets that provide a high caloric intake in a small volume, since sweets-eating behavior is one of the more common reasons for failure of gastric restriction procedures for obesity. We have routinely excluded patients with identifiable sweets-eating behaviors from undergoing gastric restrictive procedures. Despite this, we have seen many patients develop sweets-eating behavior when faced with the postoperative limitations in quantity of oral consumption imposed by the procedure. Repeated dietary counseling may help avoid, and occasionally treat, such oral indiscretions. However, more commonly, patients who act on their cravings for sweets either fail to lose weight or regain their lost weight.

While few surgeons perform routine contrast studies of the gastrointestinal tract before band adjustment, this was mandated in the 'A Trial' in the United States. Band stomal diameter was estimated radiologically before and after band adjustment. A representative contrast study of a well-positioned Lap-Band is seen in Figure 7.4. A contrast study of a patient with acute gastric herniation through the band with obstruction is depicted in Figure 7.5. In our series of 36 Lap-Bands, 18 of 25 patients (71%) who had both preoperative and postoperative contrast studies demonstrated postoperative esophageal diameter an average of 182 % (range 100 to 286%) of baseline diameter over an average period of 21 months. Although there are no standard methods in the literature for measuring esophageal dilation by contrast esophagram, we standardized our measurements by using the vertebral body height and band diameter to provide internal controls for variable film magnification. Our review suggests that normal resting esophageal diameter should be less than 16 mm. In our LASGB patients, stoma diameter and amount of weight loss did not correlate with the degree of esophageal dilatation. The mean preoperative esophageal diameter was 2.2 cm (range 1.4 to 3.1 cm), which increased to 3.3 cm (range 1.9 to 4.8 cm) postoperatively. Two patients presented to our clinic from other centers with esophageal dilatation and resulting symptoms. Saline was removed from the reservoir, which only minimally decreased the degree of dilatation. Contrast exams revealed 11 patients had delayed esophageal emptying, five had decreased motility, and seven had both. Bands were removed in two patients due to symptoms related to esophageal dilatation. Twelve of 17 patients with severe dilatation were symptomatic with dysphagia, vomiting or severe reflux. Five of six patients with the greatest postoperative esophageal dilatation ($\geq$ 200% of baseline diameter) were symptomatic. Two patients had pouch dilatation. Of the seven patients in our series who were not significantly dilated on contrast studies, six had only short-term follow-up, suggesting that dilatation may not develop until beyond one year postoperatively.

We tightened the band despite the presence of varying degrees of esophageal dilatation on the pre-adjustment contrast esophagram in eight patients. This occurred before we recognized the high incidence of esophageal dilatation in our LASGB patients and altered our adjustment strategy. Fluid was not removed from the band in several patients because of inadequate weight loss. Worsening esophageal dilatation and inadequate weight loss mandates conversion to proximal gastric bypass.

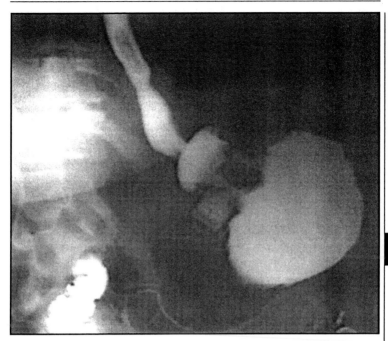

7

Fig. 7.4. Radiograph demonstrating normal positioning of the Lap-Band device.

Other patients with dilatation are being followed by interval contrast radiography to assess for regression or possible progression of the esophageal dilatation as they do not desire surgery and/or feel their weight loss is adequate.

Esophageal dysmotility and dilation is a newly recognized complication of LASGB. Intuitively, one might expect an inverse correlation between stomal diameter and esophageal dilation, but none was found. The majority of patients in our study developed new or more severe esophageal symptoms after placement of the device. Other studies support an increase in reflux after gastric banding. Ovrebo found in a study of 17 patients with the so-called "Swedish" LASGB device that acid regurgitation and heartburn increased from approx 15% to 60% after gastric banding. Other authors report complications such as food intolerance unresponsive to band deflation being attributable to pouch dilatation and/or stomal stenosis. Kuzmak, although using a previous version of the current band system, showed that early postoperative contrast study document a pouch dilation rate of 6.5% which increased to 50% over a four year follow-up. Doherty et al found that 38% of patients with an adjustable silicone gastric band required hospitalization for postprandial nausea, vomiting and severe reflux. Radiographically, these patients had enlarged proximal pouches, with delayed or absent pouch emptying and severe reflux. It is unclear if esophageal dysmotility or dilation occurred concurrently with pouch dilation in any of these studies, as these variables were not reported. Perhaps

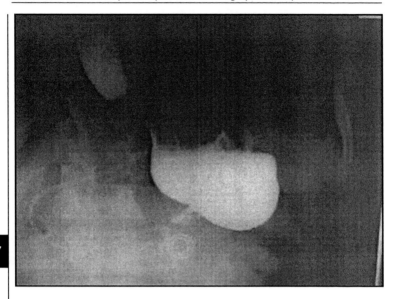

Fig. 7.5. Radiograph demonstrating gastric herniation through the Lap-Band device with obstruction.

more proximal placement of the band immediately below the gastroesophageal junction causes esophageal dilation, whereas more distal placement on the proximal stomach causes pouch dilation followed by esophageal dilation over time.

The long-term risks of esophageal dilatation are unknown but could include achalasia-like symptoms, esophageal pulsion diverticulae, or progressive development of a sigmoid esophagus that may not respond to band decompression. Appropriate management of this problem is not clear. Long-term follow-up will be required. We believe that all patients should undergo routine contrast studies at three years after device insertion. Management of the progressively dilating esophagus should include deflation of the device, despite the fact that weight re-gain is likely. Failure of the esophageal contour to return to normal should probably be treated by band removal. We have found most of our patients to have significant concern about the possible long-term health affects of esophageal dilatation, despite the lack of data on this topic. We advise such patients to undergo conversion to proximal gastric bypass.

Outcomes of LASGB

Kuzmak developed the concept of silicone gastric banding in the 1980s. He performed procedures using open surgical techniques for band placement in 311 severely obese patients. In 1986, the silicone band he used was modified to include an adjustable portion that enhanced weight loss. He reported that 57% of his adjustable banding patients achieved > 60% reduction in excess body weight in 36-month follow-up.

Kuzmak's adjustable silicone gastric band is currently called the Lap-Band (Fig. 7.1) manufactured by BioEnterics, Corp (Carpinteria, California). Another adjustable band referred to as the Swedish Adjustable Gastric Band (SAGB, AB Obtech) was developed in the 1980s in Sweden. Forsell and Hellers reported 4-year follow-up in 50 patients in whom the SAGB was placed via laparotomy. Body mass index (BMI) decreased from 46 to 27.5 kg/m^2 with a mean weight loss of 80 kg. However, other authors using this device have been disappointed with the results, particularly due to the need for reoperation in 35% of patients, with band erosion and erosive esophagitis being the most common reason for surgical revision. Other reasons for reoperation in this series included pouch dilatation, invagination of distal gastric wall through the band, leakage from the balloon, and patient dissatisfaction. When questioned two years postoperatively more than half of the patients reported vomiting, heartburn and regurgitation but 78% still pronounced themselves satisfied with the operation. Esophagitis was found in 56% of the patients at gastroscopy after two years.

Belachev and colleagues began performing adjustable banding procedures in 1991 in Belgium. They reported a favorable comparison of 200 open ASB to 210 open vertical banded gastroplasties. They subsequently performed the first laparoscopic placement of the band in 1993. Their series of 350 patients after laparoscopic implantation included patients weighing between 92 and 200 kg (mean 118 kg) with a mean BMI of 43 kg/m^2 (range 35-65 kg/m^2). The overall open conversion rate was a low 1.4%, a tribute to the authors' laparoscopic surgical expertise. Technical recommendations were made to decrease complications of pouch dilatation and gastric prolapse through the band by creating a very small proximal pouch and placing gastro-gastric sutures anteriorly to secure the band in position.

Numerous authors have now reported series of Lap-Band patients with outcome data. Most of these reports suggest that the procedure can be done with laparoscopic techniques in a high percentage of cases both safely and expeditiously as the surgeon gains the necessary technical experience over time. A few series are worth noting in detail as representative of the best available literature on the procedure.

Fielding and colleagues from Australia reported their results in 335 patients undergoing the Lap-Band procedure. Gastric herniation through the band, alternatively referred to as 'slippage', occurred in 12 patients requiring reoperation. Five bands were removed due to reflux symptoms or food intolerance. One late gastric perforation in the fundus mandated band removal. Weight loss in 58 patients followed for 18 months postoperatively was 62% of excess weight.

O'Brien and colleagues, also from Australia, evaluated the Lap-Band adjustable gastric banding system prospectively in a consecutive series of 302 patients, with data on perioperative outcome and weight loss pattern up to 4 years. Three hundred two patients (89% women; mean age 39 years, mean weight 124 kg) were entered into the study. Laparoscopic placement was used in 277 patients. The incidence of significant early complications was 4% and included two perforations of the stomach after open placement. The mean length of stay after laparoscopic placement was 3.9 days and only one complication (infected reservoir site) occurred in these patients. The principal late complication of prolapse of the stomach through the band occurred in 27 patients (9%). Significant modification of technique and patient care has enabled reduction of this complication in the latter part of the series. Mean (S.D.) excess weight loss was 51.0(17)% at 12 months (n = 120), 58.3(20)% at 24 months (n = 43), 61.6(2)% at 36 months (n = 25) and 682(21)% at 48 months (n = 12). The Lap-Band was found to be an effective method for achieving good weight loss in the morbidly obese at up to 4 years of follow-up. Laparoscopic placement has been associated with a short length of stay and a low frequency of complications. The authors suggest that the ability to adjust the setting of the device to achieve different degrees of gastric restriction resulted in progressive weight loss throughout the period of study.

We performed 36 successful Lap-Band insertions as part of a clinical trial of the device in the United States under an Investigational Device Exemption from the Food and Drug Administration between March of 1996 and May of 1998. One patient suffered intraoperative perforation of the stomach during the dissection and underwent open conversion and proximal gastric bypass since it was unsafe to place the band over the gastric repair site. Patients have been followed up to 4.5 years. Five patients (14%) have been lost to follow-up for two years or more from the time of surgery.

The average age of operated patients was 39 years (range 23- to 53 years old). Only three patients were male. Eight patients (22%) were African-American while the remainder were Caucasian. The average preoperative BMI was 44.5 kg/m² which decreased to 35.8 kg/m² at 36 months of follow-up. Average weight loss per patient was 18.4 kg, and average excess body weight lost (%EWL) was 38.4% at 36 months.

LASGBs were removed in 15 (41%) patients' 10 days to 42 months postoperatively. The most common reason for removal was inadequate weight loss in the presence of a functioning band (n = 6). This group lost a mean of only 21% of excess weight between 18 and 37 months postoperatively. The primary reasons for removal in others were: infection,[2] leakage of injected saline from the device causing inadequate weight loss,[2] or band slippage.[3] In both cases of band leakage, the leak was ultimately proven to be in the gastric portion of the device rather than in the tubing. Injury to the device at the time of surgical implantation as a result of needle puncture (suggested by analysis of the explanted band device in one case) or other errors in surgical technique may have been the cause of these band leaks. Thus, band removal rather than reservoir or tubing replacement/ repair was necessary. The patients with band slippage had concomitant poor weight loss (11-23%).

African-American patients demonstrated poor weight loss following LASGB as compared to Caucasians. The two racial groups demonstrated no significant differences in preoperative body weight, percent of ideal weight, or in body mass index. However, the postoperative percentage decrease in excess body weight and weight lost in kilograms were significantly less in African-Americans at 12, 24, and 36 months postoperatively.

Of the 15 patients undergoing band removal, four requested only simple device removal while 11 were converted to proximal gastric bypass. Of the remaining 21 patients with bands in situ, five (14%) were lost to follow-up for two or more years, but at last available follow-up (3-18 months post-op) had only achieved 5%-38% excess weight loss. Overall, six patients have requested removal of the band and conversion to GBP for inadequate weight loss in the presence of a functioning band.

Six additional patients have persistent severe obesity (BMI $\geq$35) at least two years post-op, but either refuse further surgery or are satisfied with the results.

Overall only four patients (11%) in our series achieved BMI <35 and/or at least 50% reduction in excess weight without complications. Two patients without complications have lost 79% and 85% of excess weight, where they remain at 113% and 117% of ideal weight, respectively. The overall need for band removal/conversion to GBP in our series will ultimately exceed 50%.

Weight loss in our series was poor. Average percent excess weight loss over three years in patients with intact bands was only 38%, despite the fact that a number of patients were deleted from the follow-up pool because of conversion to gastric bypass or band removal. There was no apparent correlation between stomal diameter and weight loss. Studies from other countries, however, report better results. For example, a study by Lise of 111 patients report a reduction in BMI from 46.4 to 33.1 kg/m^2 at two years follow up.

At the FDA Advisory Panel session[23] only 115 patients had been followed for three years following the procedure. Patients lost approximately one-third of excess weight and one-third of patients required either revision or removal of the device. In a series of international patients presented from Europe and Australia, surgical revision or repair of tubing or removal of the device has been necessary in 28%, the most common problem being prolapse. The mean % EWL following the procedure was 38%. In these patients, diabetes, sleep apnea, and hypertension resolved in only 22%, 40% and 55% respectively. In the Swedish Obese Subjects study,[11] most patients had either vertical-banded gastroplasty or implantation of the so-called Swedish band (which differs from the current device in design, introduction method, and reported results) and had only 23% loss of body weight despite a rather low average preoperative BMI of 42 kg/m^2. The authors also noted a rather disappointing 47% reduction in diabetes and 42% reduction in hypertension. Not surprisingly, their data show that the risk of retaining co-morbid conditions of dyspnea, chest pain, and physical inactivity, which persisted in 19%, 4%, and 17% of postoperative patients, respectively, decreased progressively with the degree of weight loss.

Summary

While technically relatively easy to perform this procedure, the overall weight reduction with the LASGB device is significantly less as compared to proximal gastric bypass. A procedure in which patients on average lose much less than half of their excess weight will not produce the dramatic resolution of co-morbidities seen with gastric bypass. Thus, significant co-morbidity such as diabetes and hypertension might be viewed as a relative contraindication to LASGB. An alternative viewpoint would suggest that improvement in co-morbidities in under half of patients undergoing this procedure is superior to no improvement in co-morbidities in patients who would decline superior forms of surgical intervention such as proximal gastric bypass.

Our results demonstrate that the LASGB procedure is not an efficacious procedure for treatment of morbid obesity in our population of American patients—and it is unclear why these results differ from those seen in other countries around the world. Furthermore, cultural, dietary, genetic, and metabolic factors might be implicated in the dramatic difference between weight loss in this American series as compared to studies around the world. Alternatively, rigorous follow-up issues, compliance issues, or even dissatisfaction of American patients with the lesser degree of weight loss achieved following LASGB might also explain the high rate of band removal and conversion to gastric bypass.

We believe the LASGB procedure must demonstrate satisfactory long-term outcomes, particularly in terms of eliminating co-morbid conditions, before the device is endorsed by the FDA as safe and effective for treatment of the estimated 10 million plus morbidly obese Americans who qualify for surgical treatment. Furthermore, current data strongly question its use in patients with preoperative gastroesophageal reflux, type II diabetes mellitus, hypertension, addiction to sweets, and, in the current study, in African-Americans. More study is required to determine long-term efficacy of the LASGB procedure in American patients.

Suggested Readings

1. Belachew M, Legrand M, Vincent V, et al. Laparoscopic adjustable gastric banding. World J Surg 1998; 22:955-963.
2. Cadiere GB, Himpens J, Vertruyen M, et al. Laparoscopic gastroplasty (adjustable silicone gastric banding). Semin Laparosc Surg 2000; 7:55-65.
3. Chelala E, Cadiere GB, Favretti F, et al. Conversions and complication in 185 laparoscopic adjustable silicone gastric banding cases. Surg Endos 1997; 11:268-271.
4. DeMaria EJ, Sugerman HJ, Meador JG, et al. High failure rate after laparoscopic adjustable gastric banding for treatment of morbid obesity. Ann Surg 2001; 233:809-818.
5. Doherty C, Maher JW, Heitshusen DS. Prospective investigation of complications, reoperations, and sustained weight loss with an adjustable gastric banding device for treatment of morbid obesity. J Gastrointest Surg 1998; 2(1):102-108.
6. Favretti F, Cadiere GB, Segato G, et al. Laparoscopic adjustable silicone gastric banding (Lap-Band®): How to avoid complications. Obes. Surg. 1997; 7:352-358.
7. Kuzmak LI. Stoma adjustable silicone gastric banding. Problems Gen Surg 1992; 9:298-317.
8. Kuzmak LI. A review of seven years' experience with silicone gastric banding. Obes Surg 1991; 1:403-408.
9. Lise M, Favretti F, Belluco C et al. Stoma adjustable gastric banding: results in 111 consecutive patients. Obes Surg 1994; 4:274-278.
10. Morino M, Toppino M, Garrone C. Disappointing long-term results of laparoscopic adjustable silicone gastric banding. B J Surg 1997; 84:868-869.
11. O'Brien PE, Brown WA, Smith A, et al. Prospective study of a laparoscopically placed, adjustable gastric band in the treatment of morbid obesity. Br J Surg 1999; 86:113-118.
12. Overbo KK, Hatlebakk JG, Viste A, et al. Gastroesophageal reflux in morbidly obese patients treated with gastric banding or vertical banded gastroplasty. Ann Surg 1999; 228:51-58.
13. Sugerman HJ, Starkey JV, Birkenhauer R. A randomized prospective trial of gastric bypass versus vertical banded gastroplasty for morbid obesity and their effects on sweets versus non-sweets eaters. Ann Surg 1987; 205:613-624.
14. Sugerman HJ, Londrey GL, Kellum JM, et al. Weight loss after vertical banded gastroplasty and Roux-en Y gastric bypass for morbid obesity with selective versus random assignment. Am J Surg 1989; 157:93-102.
15. Szucs RA, Turner MA, Kellum JM, et al. Adjustable laparoscopic gastric band for the treatment of morbid obesity: radiologic evaluation. Amer J Radiol 1998; 170:1-4.

7

16. Transcript of Proceedings of the Department of Health and Human Services Food and Drug Administration Center for Drug Evaluation and Research, Gastroenterology and Urology Panel of the Medical Devices, Advisory Committee, Monday June 19, 2000.

7

CHAPTER 8

Laparoscopic Vertical Banded Gastroplasty

Adolfo Z. Fernandez and Eric J. DeMaria

Introduction

The evolution of morbid obesity surgery has been directed by an effort to minimize complications while improving weight loss and reducing obesity comorbidities. Two different types of procedures have developed, simple and complex. Simple procedures include all restrictive procedures like the horizontal gastroplasty (HG), the vertical banded gastroplasty (VBG), or the adjustable ring gastroplasty. Complex procedures entail those that bypass segments of the gastrointestinal tract like the jejunoileal bypass, biliopancreatic diversion and the gastric bypass procedure (GBP). The complex procedure obtains better weight loss at the cost of more complications. Of all these, the most commonly performed are the GBP and the VBG—the current gold standards of bariatric surgery in their respective classes.

Surgeons will argue the superiority of GBP versus VBG and vice versa. VBG does not produce as much excess weight loss as the GBP, nor is it as durable. VBG does produce adequate weight loss improving the patient's obesity comorbidities. VBG is technically simpler to perform, offers less risk of micronutritional deficiencies, and maintains the continuity of the gastrointestinal tract. Furthermore, the VBG is considered a less morbid procedure, though some will argue to the contrary. The major disadvantage is that patients can regain the weight lost with the VBG by simple alterations in their diet. Foods that are high in fat and carbohydrates while low in volume are the main culprits. These same foods tend to cause a dumping syndrome in GBP patients and "protect" them from regaining unwanted weight. Overall, the choice between which procedure to perform depends on the surgeon's experience and comfort with the procedure.

Vertical Banded Gastroplasty

The horizontal gastroplasty (Fig. 8.1) was introduced by Printen in 1973. The initial procedure failed to produce significant weight loss because of a large pouch and easily dilated stoma. In 1979 Gomez published a modification of the horizontal gastroplasty with a smaller pouch and reinforced stoma. This modification did not achieve much success. The stoma obstructed early and often. The reinforcement stitch occasionally eroded into the lumen leading to stoma dilatation and weight gain. The horizontal gastroplasty failed to achieve adequate weight loss and reduction of any of the obesity-related morbidities.

In 1976, Tretbar first introduced the concept of a vertical pouch. He named this procedure the fundal exclusion. In this procedure, the fundus, the most compliant and expandable portion of the stomach, was excluded from the satiety process. The

Laparoscopic Bariatric Surgery: Techniques and Outcomes, edited by Eric J. DeMaria, Rifat Latifi and Harvey J. Sugerman. ©2002 Landes Bioscience.

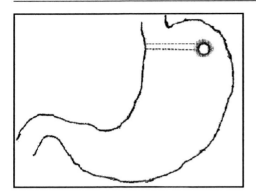

Fig. 8.1. Horizontal gastroplasty as described by Mason in 1980. The window was hand-sewn.

procedure had some success, but the success was limited by the pouch and stoma size. The prior experience with the horizontal gastroplasty and the fundal exclusion led to the development of the VBG. Mason first described it in 1982 (Fig. 8.2). Since then, multiple modifications have been reported including variations in the band, stoma size, pouch size, and stomach partition. Many feel that slight variations in these can affect the amount of weight lost or the morbidity of the procedure.

8

Techniques

The three most important components of the VBG are the band, the pouch and the stomach partition. Many authors have varied the size and composition of the band. Gore-Tex, Marlex, silastic, silicone, and adjustable bands have been used with varying degrees of success. Suter found that there was a significantly higher incidence of complications with silastic and adjustable bands when compared with Marlex mesh bands. The latter two tended to produce more obstructive type complications requiring re-operation. Ashley had a greater rate of stenosis and vomiting with Gore-Tex bands, compared to Marlex bands. One confounding variable was that the Gore-Tex bands were smaller than the Marlex bands. Naslund used a silicone band and had a 31% rate of complications including stomal stenosis, esophagitis, band erosion and staple line breakdown. Mason exclusively used the Marlex bands stating that the Marlex became incorporated into the neck of the VBG, thus preventing any slippage or intraluminal migration. The Marlex mesh bands produced the least amount of complications and were the most popular choice.

Sizing of the bands varied from author to author. Mason looked at three different sizes of the Marlex band—4.5 cm, 5 cm and 5.5 cm. Of the three, the 4.5 cm band had the best five year success in weight loss (defined as loss of more than 25% of the excess weight) but the highest five year reoperative rate (9%). Patients also tended to have stricter dietary restrictions and more difficulty tolerating a regular diet leading to dissatisfaction. Many then began to take in higher calorie drinks and processed food. Ultimately, this dietary change led to weight gain. The 5.0 cm band group did the best overall with over 78% of patients having successful weight loss while only 6.1% required re-operation. Most importantly, these patients tolerated regular food better and were more compliant with the postgastroplasty diets. The 5.5 cm bands had disappointing weight loss results. MacLean used a double layer of

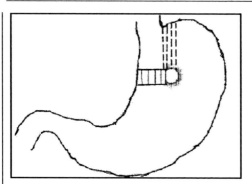

Fig. 8.2. Vertical banded gastroplasty as described by Mason.

Marlex with a circumference of only 4.3 to 4.5 cm. He experienced high rates of band migration into the gastric lumen (25%) and staple line disruption (48%). Other authors used internal calibration to determine the size of the stoma. Naslund used a 32 F gastric tube to calibrate his band. Out of 198 patients, he reported only one stomal stenosis and 19 staple line disruptions. Mason was the only author to critically evaluate outcomes of different band sizes, and Suter published the only comparison of band type. Overall, the 5cm Marlex band was the technique fraught with the least amount of complications in the literature.

Pouch size is another disputed area of VBG. The pouch size recommended by Mason was on average about 15 cc under 70 cm of water pressure. Naslund reported that there was no difference in weight loss whether the pouch volume was 20 ml, 30 ml or 40 ml. He did suggest that patients with pouches 40 ml or greater did worse than those with smaller pouches. Ashley's median pouch volume was 20 mL, ranging from 15-50 ml. No one study evaluated the impact of pouch size on weight loss or complications. The data was unable to show any differences in weight loss between pouch sizes, but the consensus between authors was to keep the pouch as small as possible.

The creation of the stomach partition also has gone through multiple changes from the two row staplers to the four row staplers. Currently the best results are found with either a four-row stapled partition or complete division of the stomach. The main difference is that should the staple line dehisce, then either the patient regains weight because of the loss of the partition or gets a life-threatening gastric leak. Toppino found a 30.5% rate of staple line dehiscence in patients who had initial success but who later failed to maintain the weight loss or regained weight. He used both the stapled partition technique and complete division. He did not specify if one technique failed more often than another. Suter had an 11% incidence of staple line dehiscence using varied staplers (simple TA90, double TA90, TA90 B) but never divided the stomach. Naslund reported 19 patients of 70 (27%) with a staple line disruption. He used the TA90 stapler twice or the TA90 B once. The evolution of the partitioning method has mainly relied on the development of newer and better stapling devices. No one study has compared the stapled partition to the divided pouch.

Outcomes

The combination of these elements produces a gastric restrictive procedure aimed at weight loss and reduction of obesity-linked morbidities. The long-term results have been mixed (Table 8.1). Mason has the largest running series. His series consisted of 1,298 patients from February 1981 to June 1988. The risk of leakage and peritonitis was 0.6% and mortality was 0.24%. Of the 1,298 patients, 759 underwent the VBG with 5 cm Marlex band (VBG5). The reoperation rate for this group was 1.4% each year for the first five years. In this group 313 patients completed five years of follow-up. At five years, the morbidly obese (MO) fraction (those less than 225% of ideal body weight (IBW)) lost 51.0% of their excess weight. The super obese (SO) fraction (those greater than 225% of ideal body weight (IBW)) lost 43.8% of their excess weight. Baltasar reported similar results in his first 100 patients undergoing VBG after a minimum of five-year follow-up. The percent of excess weight loss (%EWL) at five years was 54.3% in 84 patients. Suter's %EWL remained between 50-60% in his cohort of patients up to nine years postoperatively. Overall, the data suggests that the patients lose about 50% of their excess weight with resolution of their comorbidities. In our experience, the Roux-en-Y GBP produced a significantly greater %EWL than the VBG. Headley, however, showed that despite the greater weight loss and greater satisfaction within the GBP group, the overall improvements in the patients' hypertension, diabetes mellitus and orthopedic problems were "similar."

8

Despite the greater %EWL with the GBP, the VBG is reportedly less morbid. The major difference involves the complications associated with bypass of the duodenum and part of the jejunum. This produces micro-nutritional deficiencies. Iron, calcium, and folate are the principal ones that require oral replacement. Furthermore, the bypassed section of the gastrointestinal tract is not easily evaluated endoscopically or radiographically. The consumption of concentrated sweets or milk products can also produce a dumping syndrome. Because the patient feels terrible after sweets, the dumping syndrome is actually an advantage for the GBP. In our experience, the GBP has been much more successful with sweet eaters than the VBG. When comparing perioperative complications, there is little difference between the groups. On the average the mortality of both procedures is less than 1%. The rates of complications range from 21% to 44% for the VBG. The rates of gastric leak and wound infection are very low, ranging from 0.6% to 5.7% for gastric leak and 1.4% to 7.6% for wound infection. More common complications are those of stomal stenosis, vomiting and dilatation of the pouch. Their incidence ranges from 6.5% to 20%; the majority require reoperation. These rates of complications are comparable to the GBP.

Overall, the morbidity and mortality of the GBP and the VBG are not significantly different. VBG has some clear advantages. They are (1) maintenance of the GI tract continuity, (2) ease of reversibility, (3) little metabolic side effects, and (4) fairly reproducible results. On the other hand, its disadvantages include (1) a high reoperation rate, (2) smaller %EWL, (3) fewer diet requirement, and (4) poor longevity. The decision of which procedure to use is partially based an the surgeon's training and comfort level with a certain procedure.

Table 8.1. **Results of the larger series of VBG reported in the literature.**
Of note, Mason reported results for over 1,000 patients; these
results represent the VBG5 subgroup.

Author	Number of Patients	Mortality Rate	Reoperation Rate	Leak Rate	% of Excess Weight Loss
Mason (1992)	313	0.24%	6.1%	0.6%	MO 58.8% SO 51.6%
Ashley (1993)	114	0%	N/A	2.6%	46.4%
Naslund (1997)	198	0.5%	44%	5.7%	N/A
Baltasar (1998)	100	1%	25%	4%	54.3%
Suter (2000)	197	0%	29.4%	N/A	50-60%

Laparoscopic Vertical Banded Gastroplasty

The evolution of obesity surgery to laparoscopy was powered by the advantages offered by laparoscopic techniques. These included decreased pulmonary complications, reduced analgesic requirement, quicker patient mobilization, shorter hospital stays, shorter recovery periods, and decreased incidence of incisional hernias. Minimally invasive surgery provides significant benefits, especially in the field of obesity surgery. Obviously the complexity of the procedures and the patient's body habitus requires development of specialized instruments and new techniques to facilitate successful completion. The laparoscopic VBG (LVBG) was first described by Chua and Mendiola in 1995. It is still evolving and changing.

Technique

The performance of the LVBG is technically demanding requiring advanced laparoscopic skills. The procedure itself requires up to five ports. They are usually placed in the right upper quadrant, inferior to the xiphoid, above the umbilicus, and at the left subcostal margin (2). The gastrophrenic and gastrohepatic ligaments are opened. The retrogastric space is opened and dissected free. A window is created at the angle of His. An orogastric tube is then passed into the stomach and directed along the lesser curve. This helps determine the size of the pouch and its stoma. Next a gastric window is created 3 cm to the left of the lesser curve and 4 to 7 cm from the window at the angle of His. A circular stapler is inserted via the subxiphoid port and used to create the gastric window. A linear stapler, either cutting or non-cutting is inserted through the gastric window up to the angle of His and fired. Once the pouch is created, the band is inserted and secured into position. The band is either marked extracorporeally at a certain length or adjusted snugly around the orogastric tube. The band is secured to itself by nonabsorbable stitches and the procedure is complete.

Many slight variations to this technique exist. Some authors perform this procedure hand assisted (HLVBG). The advantage of the hand-assisted technique is that the learning curve and technical demand is reduced. The procedure can be done in a shorter period of time while retaining the advantages of the laparoscopic tech-

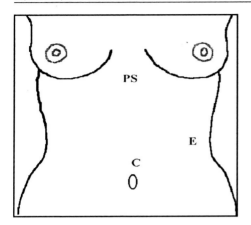

Fig. 8.3. Port placement for the hand-assisted laparoscopic VBG as described by Gerhart. PS = Pneumo Sleeve, C = Camera and E = 33 mm port for circular stapler.

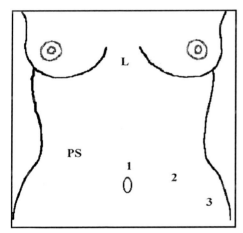

Fig. 8.4. Port placement for the hand-assisted laparoscopic VBG as described by Watson and Game. PS = Pneumo Sleeve, L = Liver retractor, 1 = 11 mm port, 2 = 11 mm port and 3 = 5 mm port.

8

nique. The disadvantage is that it requires the hand to be inserted into the abdomen, just inferior to the xiphoid, thereby necessitating a larger incision. This creates a risk of incisional hernia at the site. Two or three other ports are used—a supraumbilical camera port and one to two left subcostal working ports (Figs. 8.3 and 8.4). The initial experience with this technique is quite small.

Another variation of the LVBG is Cagigas' "no punch" technique (Fig. 8.5). His group does not create a gastric window. Instead they use a cutting, reticulating linear stapler to create the gastric pouch by starting at the angle of His. At the base of the pouch, they place a Gore-Tex band. The stoma size is calibrated around a gastric tube that is placed prior to the division of the gastric pouch. The advantage of this method is that the procedure can be performed through 10 mm and 12 mm ports, since the large circular stapler is no longer required (Fig. 8.6). A disadvantage is that

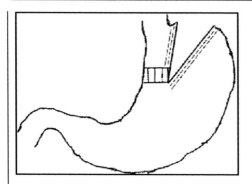

Fig. 8.5. A schematic representation of the "no punch" technique laparoscopic VBG described by Cagigas.

the stomach must be divided placing the patient at risk for peritonitis should he have a staple line dehiscence. The preliminary results have yet to be reported.

Outcomes

The reported series of the LVBG and HLVBG have small group numbers and do not have long term follow-up. Gerhart reported his results of a small series of 26 patients who underwent HLVBG with follow-up ranging 3 to 18 months. The %EWL at 12 months was 52.7%, which is lower than the %EWL seen in the open procedure at one year. So far he has experienced three minor wound infections, two patients with clinically significant atelectasis, three incisional hernias, one leak, one pouch outlet obstruction, and one staple line dehiscence. The rate of complications is 42% and rate of reoperation is 24%. Gerhart results are similar to those seen with the VBG.

Bleier reported his initial experience with the HLVBG in 46 patients. He compared them with 46 historical controls (VBG) and found that the HLVBG patients ambulated and ate earlier. The operation was quicker, and ICU stay was shorter. Pulmonary and wound complications were reduced compared to the open group. Pain was more easily managed within the HLVBG group. The one significant difference was three gastric leaks (6.5%) in the HLVBG group but none in the open group. Two of the leaks were managed conservatively and resolved spontaneously. The third required laparotomy and repair. This patient remained in hospital for 96 days. Excluding this patient from the average hospital length of stay (LOS), the HLVBG group had a significantly shorter LOS, 4.3 versus 7.7 days. Overall, the patients faired better in the HLVBG group although the rates of gastric leak were increased. These results are the expected benefit of the laparoscopic technique. The only concern is the high rate of gastric leak. With Bleier's divided gastric pouch. Staple line disruption has been seen in up to 40% of cases with non-divided gastric pouches. This becomes a greater concern in cases where the pouch is divided.

The experience with the LVBG is somewhat more extensive than with the HLVBG. The initial results are as expected. Lonroth attempted 38 LVBG's and successfully completed 35. Three of the patients had to be reoperated for a retained and transected gastric tube, another for a staple line leak, and the last for persistent fever with a subphrenic fluid collection. On the other hand, the LVBG group faired better than the open controls in that they had less pain, quicker mobilization and

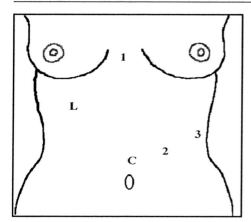

Fig. 8.6. Port placement for the "no punch" technique laparoscopic VBG described by Cagigas. L = Liver retractor, C = Camera, and 1-3 = 10-12 mm working ports.

less respiratory function impairment. Azagra had similar early results. He prospectively randomized 68 patients to either LVBG or an open VBG. The LVBG procedure took significantly longer, 150 minutes versus 60 minutes. The laparoscopic group had one intraoperative complication (bleeding) requiring conversion to open. The open group did not have any. Postoperatively, the laparoscopic group faired better overall. Both groups had gastric leaks, one in LVBG and two in VBG. The LVBG group had fewer wound infections, one versus four, and no incisional hernias. The open group had six incisional hernias in 38 patients. Davila-Cervantes also compared two well-matched groups of 20 patients who had been the first to undergo VBG and LVBG at his institution. There were no leaks in either group. No other significant complications were reported.

These studies showed that the LVBG could be performed safely. There is of course a learning curve. Some of the authors noted that most of their complications occurred early on in their experience. The laparoscopic approach was effective in reducing postoperative pain, respiratory and wound complications. It also reduced the hospital stays in some of the series. Furthermore, it was noted that reoperation in the LVBG was possible laparoscopically without added risk. We have routinely reoperated on our laparoscopic gastric bypasses laparoscopically with good success. Larger series are still required to establish if the long-term results will be comparable to those of the open VBG.

The literature is lacking in publications with long term follow-up to determine LVBG's effectiveness. The longest follow-up period to date was 36 months in Naslund's series. He started with 60 patients, 15 of whom were converted to open VBG and used as controls. At 36 months there was no statistical difference between the absolute weights of either group. This result is deceiving in that less than one-third of the LVBG patients had been followed for 36 months and the open group had a significantly ($p=0.001$) greater mean body mass index (BMI) preoperatively, 48.8 kg/m^2 versus 41.9 kg/m^2 respectively. Davila-Cervantes showed at 12 months the %EWL was slightly higher in the open group, 57% versus 52%. These initial results indicate that the LVBG may not produce as much excess weight loss as the open VBG early on. Though these findings are preliminary, they are concerning.

8

Long term follow-up with well-matched controls is needed to establish the weight loss potential for the LVBG.

Conclusion

Vertical banded gastroplasty has been established as a reliable option in the surgical treatment of morbid obesity. Like all procedures, it has its limitations especially in obese patients who are also "sweet eaters". Mason and others have shown that the VBG has a reliable weight loss result and produces good resolution of the obesity comorbidities. The major drawbacks include the possible production of significant dietary alterations. Some patients may be limited in the types of foods that they can ingest thus resulting in dissatisfaction with the procedure's result. This can lead to alterations of the diet producing the cessation of weight loss or worse weight gain. The advantages are that the gastrointestinal tract is maintained in continuity and metabolic complications are minimal. The efficacy of the laparoscopic vertical banded gastroplasty is yet to be determined. The early results are promising, but more work is required to further refine the procedure.

Selected Readings

1.   Ashley S, Bird DL, Sugden G et al. Vertical banded gastroplasty for the treatment of morbid obesity. Br J Surg 1993; 80: 1421-1423.
2.   Azagra JS, Goergen M, Ansay J et al. Laparoscopic gastric reduction surgery. Surg Endosc 1999; 13: 555-558
3.   Baltasar A, Bou R, Arlandis F et al. Vertical banded gastroplasty at more than 5 years. Obes Surg 1998; 8: 29-34.
4.   Bleier JI, Krupnick AS, Kreisel D et al. Hand-assisted laparoscopic vertical banded gastroplasty. Surg Endosc 2000; 14: 902-907
5.   Cagigas JC, Martino E, Ingelmo A et al. "No punch" technique of laparoscopic vertical banded gastroplasty for morbid obesity. Obes Surg 1999; 9: 407-409.
6.   Davila-Cervantes A, Ganci-Cerrud G, Gamino R et al. Open versus laparoscopic vertical banded gastroplasty: a case controlled study with 1-year follow-up. Obes Surg 2000; 10: 409-412
7.   Gerhart CD. Hand-assisted laparoscopic vertical banded gastroplasty. Arch Surg 2000; 135: 795-798
8.   Gomez CA. Gastroplasty in morbid obesity. Surg Clin N Am 1979; 59: 1113-1120
9.   Headley WM (1992) Gastric bypass versus vertical banded gastroplasty. Prob Gen Surg 9: 332-344.
10. Knol JA, Strodel WE, Eckhauser FE (1987) Critical appraisal of horizontal gastroplasty. Am J Surg 153: 256-261.
11. Lonroth H, Dalenback J, Haglind E, Josefsson K, Olbe L, Olsen F, Lundell L (1996) Vertical banded gastroplasty by laparoscopic technique in the treatment of morbid obesity. Surg Lap Endosc 6: 102-107.
12. Lonroth H, Dalenback J. Other laparoscopic bariatric procedures. World J Surg 1998; 22: 964-968.
13. Mason EE. Vertical banded gastroplasty for obesity. Arch Surg 1982; 117: 701-706
14. Mason EE, Maher JW, Scott DH. Ten years of vertical banded gastroplasty for severe obesity. Prob Gen Surg 1992; 9: 280-289.
15. Mason EE, Tang S, Renquist KE. A decade of change in morbid obesity surgery. Obes Surg 1992; 7: 189-197.
16. Mason EE. Gastric surgery for morbid obesity. Surg Clin N Am 1992; 72: 501-513.

17. Mason EE, Doherty C, Cullen JJ et al. Vertical gastroplasty: evolution of vertical banded gastroplasty. World J Surg 1998; 22: 919-924.

18. Naslund E, Backman L, Granstrom L et al. Seven-year results of vertical banded gastroplasty for morbid obesity. Eur J Surg 1997; 163: 281-286.

19. Naslund E, Freedman J, Lagergren J et al. Three-year results of laparoscopic vertical banded gastroplasty. Obes Surg 1999; 9: 369-373.

20. Natalini G, Breccolotto F, Carloni G et al. Laparoscopic adjustable vertical banded gastroplasty: a new method for treatment of morbid obesity: preliminary experience. Obes Surg 1999; 9: 55-56.

21. Printen KJ, Mason EE. Gastric surgery for relief of morbid obesity. Arch Surg 1973; 106: 428.

22. Sugerman HJ, Londrey GL, Kellum JK. Weight loss with vertical banded gastroplasty and Roux-Y gastric bypass for morbid obesity with selective versus random assignment. Am J Surg 1989; 157: 93-102.

23. Sugerman HJ, Kellum JK, DeMaria EJ. Conversion of failed or complicated vertical banded gastroplasty to gastric bypass in morbid obesity. Am J Surg 1996; 171:263-269.

24. Suter M, Jayet C, Jayet A. Vertical banded gastroplasty: long term results comparing three different techniques. Obes Surg 2000;10: 41-46.

25. Toppino M, Morino M, Capuzzi P. Outcome of vertical banded gastroplasty. Obes Surg 1999; 9: 51-54.

26. Tretbar LL, Taylor TL, Sifers EC. Gastric plication for morbid obesity. J Kans Med Soc 1976; 77: 488.

27. Watson DI, Game PA. Hand-assisted laparoscopic vertical banded gastroplasty. Surg Endosc 1997; 11: 1218-1220.

8

CHAPTER 9

Hand-Assisted Laparoscopic Roux-en-Y Gastric Bypass

Michael A. Schweitzer and Eric J. DeMaria

While hand-assisted techniques have been shunned by some in the laparoscopic community, what better population to whom to apply it than the morbidly obese where conventional laparoscopic instruments may fall short. The total laparoscopic technique is difficult to learn and requires a very skilled endoscopic surgeon along with an assistant who is also well-trained in advanced laparoscopic techniques. The tactile sensation returned with hand-assisted technique allows for the operating surgeon to have more control over dissection and bowel manipulation as he/she would in open surgery but with the distinct advantage of using only a 7 cm to 8 cm incision along with a few trocar sites. The surgeons hand offers one of nature's best atraumatic retractors and clearly a superb blunt dissector. The hand assist aids in bringing the tissue closer to the dissecting, cutting and coagulation devices. This is also evident during endoscopic stapling and clipping where the tissue can be hand directed into the jaws of the stapler or clip applier while keeping away tissue not meant to be divided, anastomosed or clipped. A novice laparoscopic surgeon may be hindered by the lack of depth perception in advanced laparoscopic cases. This may be overcome by proprioceptive feedback from the intraperitoneal hand, which will allow the surgeon to have better depth perception despite a two-dimensional view through the endoscope.

Originally, hand-assisted laparoscopy involved taking a gloved hand and placing it directly into the abdomen through a tight incision the size of ones wrist. This technique was used in colon surgery to extract the bowel and then place the proximal end with the anvil, and attach it to the stapler distally. Problems arose if the incision was too long and pneumoperitoneum could not be maintained. It was also difficult to rotate and move your hand in all directions and still maintain pneumoperitoneum. Surgeons would also suffer from hand fatigue given the tight geometry they had to work in. Several companies have invented devices that allow much greater movement while maintaining pneumoperitoneum. They also will allow you to remove your hand and operate through the incision. Two devices that work in morbidly obese patients are the Pneumo Sleeve® (Lifequest, Rosewell, GA) and Handport® (Smith & Nephew, Andover, MA). The Pneumo Sleeve uses a protractor device to protect and open the incision along with a port that is temporarily glued to the skin around the incision. The sleeve is then placed on the surgeons nondominant hand and attached to the port. The Handport device, in contrast, uses a double inflatable collar that is placed in the incision to maintain pneumoperitoneum

Laparoscopic Bariatric Surgery: Techniques and Outcomes, edited by Eric J. DeMaria, Rifat Latifi and Harvey J. Sugerman. ©2002 Landes Bioscience.

and is then attached to the extra long sleeve that was placed around the surgeons nondominant hand.

Table 9.1 lists potential advantages of hand-assistance over a total laparoscopic approach. To date, we routinely use only three 12 mm trocars and three 5 mm trocars in our total laparoscopic technique that uses a linear stapler to create the gastrojejunostomy. We therefore use hand assistance for patients with umbilical or ventral hernias where a 7 to 8 cm incision would be needed anyway to close the defect. A hand inside the abdomen may also be helpful in salvaging a total laparoscopic gastric bypass, for example, in a 400 lb man with a significant hiatal hernia where a hand is helpful to retract the stomach downward to identify the angle of His and transect the stomach. The bariatric surgeon who is trying to learn laparoscopic gastric bypass may benefit early on from a hand-assist approach where the small intestine is eviscerated and the jejuno-jejunostomy is done by the usual open method. This will allow the surgeon to learn the technique of creating the gastrojejunostomy under pneumoperitoneum but still have a hand to assist.

Operative Techniques

Creation of Roux Limb and Jejuno-jejunostomy

Preoperatively on the day of surgery the patient is given 2 g of cefazolin and intermittent lower leg venous compression devices are applied on each lower extremity. The patient is placed in the supine position with the operating surgeon on the left side and the assistant on the right side of the patient (Fig. 9.1). We do not use the lithotomy position with stirrups since the risk of injury to the lower extremities may be greater due to the strain from the increased weight and size of the obese legs. A 12 mm Versaport® (US Surgical Corporation, Norwalk, CT) is placed in the left subcostal position using the Visiport® (US Surgical Corporation, Norwalk, CT) with a 10 mm zero degree laparoscope that allows direct visualization of the abdominal wall layers being incised. The abdomen is insufflated and a right upper quadrant 12 mm trocar is placed under direct vision with a 10 mm 45° endoscope.

The Pneumo Sleeve® port is then applied around the proposed midline incision site. It is important to place the port on while the abdomen is insufflated for it to adhere best and then make the hand incision. This incision is made at the superior aspect of the umbilicus if possible; however, placement of the incision will depend on the distance from the patient's xiphoid to their umbilicus. By making the incision to incorporate the umbilicus the resulting scar will be cosmetically more appealing. If the umbilicus is a long distance from the xiphoid then the incision will need to be more cephalad. The skin incision is made the same size as the operating surgeon's glove size, approximately a 7.5 cm incision with the linea alba opened for 8 cm. The fascial incision should be made more cephalad so the small bowel near the ligament of Treitz can be eviscerated through the wound. By making a midline incision, it can be easily extended into a full laparotomy incision for open gastric bypass if a problem develops. The Protractor® device is then inserted which will provide good retraction and wound protection.

The ligament of Treitz is located by tracing the small bowel proximally. An EndoGIA II® (US Surgical Corporation, Norwalk, CT) with a 60 mm cartridge and 2.5 mm staples is used to divide the jejunum approximately 25 to 75 cm from the

---

Table 9.1. Potential advantages of hand assistance

- Already need a 7 to 8 cm incision to repair an umbilical or ventral hernia that is common in this population.
- May use to salvage a total laparoscopic case that needs extra-assistance that a hand can provide.
- If a skilled assistant for a total laparoscopic approach is not available.
- The possible use in higher BMI patients.
- Inexperienced surgeon may use technique as a bridge while acquiring the skills to do the total laparoscopic approach.

---

ligament of Treitz. A penrose drain is sewn to the end of the Roux limb. A vascular EndoGIA II® with a 45 mm cartridge and 2.0mm staples is used to divide the mesentery. A jejuno-jejunostomy is performed approximately 60 cm distal to where the jejunum was divided. Enterotomy holes are made in the eviscerated segments of jejunum for a side to side anastomosis using an EndoGIA II® with a 60 mm cartridge carrying 2.5 mm staples. The enterotomy hole is then closed using a suture or staple technique with care to avoid narrowing of the lumen. The jejuno-jejunostomy may also be performed intracorporeally using hand assistance. We have found in most instances the jejunum can be eviscerated enough into the hand incision to perform the anastomosis using the endoscopic stapler which will ultimately save operative time.

9

Division of the Gastric Pouch

The Pneumo Sleeve® device is placed over the left hand, which is then inserted into the abdomen, and pneumoperitoneum is reestablished (Fig. 9.1). We have found the 45°-angled viewing laparoscope most helpful for the dissection. The left lateral segment of the liver is retracted using a Nathanson retractor through a subxiphoid 5 mm puncture made by a trocar which is removed to allow insertion of the retractor which is then held in position with a commercially available metallic arm attached to the OR table (Thompson Surgical Instruments, Inc., Traverse City, MI 49684). A 12 mm Versaport® is then placed in a subxiphoid position. A 5 mm trocar is place in the right and left lateral positions. The phrenoesophageal ligament is taken down using the 5 mm UltraShears® (US Surgical Corporation, Norwalk, CT) to identify the angle of His. The lesser omental peritoneum is taken down with the UltraShears®. Palpation of an aberrant left hepatic artery is easily done since a hand is inside. A space is created separating the lesser curve of the stomach from the neurovascular bundle near the third branch off the left gastric artery. A finger on the posterior side is used to speed this dissection and can also control bleeding (Fig. 9.2). Caution should be used to not use surgical clips where the stapler will be placed since it will not work across a clip. An EndoGIA II® with a 60 mm cartridge carrying 3.5 mm staples is fired multiple times from the lesser curve side to the angle of His to make a 15 cc to 30 cc proximal gastric pouch. The laparoscope, which has been changed to 45° scope, is switched between the two ports for optimal visualization. A finger is used throughout this stapling to aid in the dissection and to orient the stapler toward the angle of His. It is important not to staple horizontal and make too large a pouch. A vertical pouch has thicker, less distensable gastric tissue.

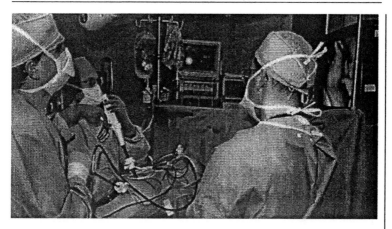

Figure 9.1. Hand assisted laparoscopic gastric bypass. Operating surgeon stands on left side and assistant on right side of the table.

9

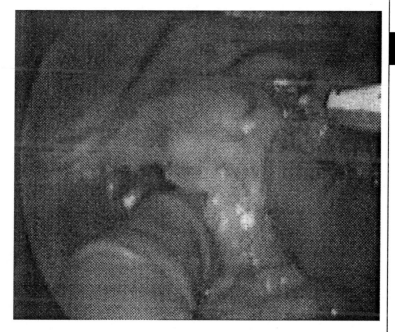

Figure 9.2.The hand helps dissect the neurovascular bundle off the lesser curve before division of the pouch.

The Roux end of the jejunum is then brought through the transverse mesocolon in a retrocolic retrogastric using finger dissection just above the ligament of trietz. The penrose attached to the Roux limb is placed into the lesser sac through this window. The patient is then placed back in steep reverse trendelenberg position (the operating table should have a foot-board attached) and the omentum and transverse colon are retracted inferiorly to visualize the lesser sac through the line of gastric transection. The penrose drain is retrieved from the lesser sac posterior to the distal stomach and the attached Roux limb is carefully pulled up into the lesser sac making certain that the limb is not inadvertently twisted.

The Linear Stapling Technique for the Gastrojejunostomy

The Roux limb is then sutured using the Endo-stitch device with 2-0 polysorb (USSC) to the gastric pouch on the lesser curve side with two interrupted stay sutures. A small enterotomy is made between the sutures in the Roux limb and a similar size gastrotomy in the pouch in order to place 35 mm of a 45 mm length cartridge with 3.5 mm staples to create the gastrojejunostomy. The anesthesia team carefully passes a 30 F bougie from the mouth through the gastrojejunal anastomosis. The bougie can be seen through the opening that was formed after the stapler was removed. Three stay sutures are again used, just as in the jejuno-jejunostomy, to elevate the tissue so that the 60 mm length cartridge with 3.5 mm staples can be used to close the openings. The stapler is brought down on top of the bougie while retracting the tissue to be transected. This firing will then form a 11 to 12 mm gastrojejunal anastomosis. The penrose drain is cut free from the Roux limb and removed from the peritoneal cavity.

EEA Stapler Technique Using Transoral Anvil on a Pull-Wire Method for the Gastrojejunostomy

A flexible endoscope is then inserted into the proximal pouch through the mouth to begin the placement of the anvil device. The scope light transilluminates a location on the gastric pouch where the anastomosis would be performed and a snare was introduced through the endoscope and visualized distending the gastric wall in the correct location. A small electrocautery opening is made with the scissors and the snare delivered into the abdomen. A 96 inch #2 looped prolene suture (or a looped wire from a percutaneous gastrostomy kit) is then placed in the open snare and the entire endoscope with attached suture material is withdrawn from the patient's mouth. A surgeon maintains control of the other end of the long wire suture for subsequent tension to pull down the anvil. A 21 mm circular stapler anvil (Ethicon Endosurgery, Cincinnati, Ohio) was then secured to the opened end of the looped wire by bringing it through an opening in the side of the anvil and looping it around the anvil's head. Tension on the wire by the surgeon's hand transabdominally, then allows the anvil to be delivered transorally to the pouch. It is important to elevate the angles of the jaw forward and deflate the endotracheal balloon with the anesthesiologist holding the endotracheal tube. It is important not to force the anvil through the cricopharyngeus muscle. If the opening where the stem of the anvil penetrates the pouch widens then it may be necessary to place a purse string suture. The end of the Roux limb is opened with the harmonic scalpel and the EEA stapler is placed through the left upper quadrant trocar after dilating the site. The stapler is place through the Roux limb and then connected to the anvil, fired, and then removed. A balloon trocar is then replaced in the dilated trocar site so that a air tight seal is formed, thereby, preventing the loss of pneumoperitoneum. The end of the Roux limb is then closed with a 2.5 mm stapling cartridge with the excess tissue and the penrose drain removed from the peritoneal cavity.

Transgastric Anvil Placement Technique for the Gastrojejunostomy

The 21 mm circular stapler anvil is delivered into the abdomen through a widen left upper quadrant port site and then a balloon trocar is used afterwards to prevent an air leak. Several methods have been described but we simply placed a #1 prolene suture with a large needle, that is partially flattened, on the anvil and open the stomach distal to where the pouch will be divided. The nasogastric tube is removed and then the suture is placed intragastric and brought out at the point the anvil should come out of the pouch. The suture is then pulled and the anvil brought through the pouch. The opening is closed with a firing of the endoscopic stapler after holding it up with two stay sutures. The pouch is then divided as previously described but instead has an anvil inside. If the opening where the stem of the anvil penetrates the pouch widens then it may be necessary to place a purse string suture. The end of the Roux limb is opened with the harmonic scalpel, and the EEA stapler is placed through the left upper quadrant trocar after dilating the site. The stapler is placed through the Roux limb and then connected to the anvil, fired, and then removed. A balloon trocar is then replaced in the dilated trocar site so that a air tight seal is formed, thereby, preventing the loss of pneumoperitoneum. The end of the Roux limb is then closed with a 2.5 mm stapling cartridge with the excess tissue and the penrose drain removed from the peritoneal cavity.

9

Completion of the Operation

At this point we gently perform an upper endoscopy to inspect the anastomosis for patency and homeostasis. The Roux limb is compressed with your fingers to prevent insufflation of the distal bowel. The anastomosis is then placed under irrigation fluid and insufflated. If a small leak is detected it can be closed with laparoscopic sutures and then retested with air insufflation. If reinforcement sutures are placed it should be done with the endoscope in the Roux limb so as not to cause narrowing of the anastomosis.

The three mesenteric defects are closed under pneumoperitenuem using interrupted sutures. A closed drain is placed in the left upper quandrant. The 8cm fascial incision is closed and all sites are injected with Marcaine®. Postoperatively, the patient is placed on low molecular weight heparin and a clear liquid diet is begun the next morning. On day two an upper gastrointestinal series is performed to rule out a leak. The patient is usually discharged on the afternoon of day two or three.

Conclusion

A hand-assisted method is technically easier for the novice laparoscopic surgeon and allows for more control when compared to pure laparoscopic methods. It can aid in the division of the small stomach pouch by retracting the stomach caudad as the angle of His approaches. It also helps in one of the most difficult parts of the case, small bowel manipulation in a morbidly obese patient. There is also the reassurance of inspecting the divided jejunal limbs during the open part of this procedure to ensure adequate blood supply. In a total laparoscopic approach subtle ischemia may not be apparent on the monitor.

Hand-assisted gastric bypass may have an improved cosmetic advantage when compared to open gastric bypass. This is especially true in a patient with an umbilical or supra-umbilical hernia that would need an incision starting at the xiphoid and going down to the umbilicus in open gastric bypass surgery.

In analysis of the prospective data collected on 25 hand-assisted gastric bypass patients at the Medical College of Virginia Hospitals costs were on average $4,444.00 higher when compared to open gastric bypass patients. Hospital stay was only slightly less compared to open (3.6 vs. 4.6 days) but not statistically significant. The incisional hernia rate through the hand incision was 20% and was therefore equivalent to the open gastric bypass patients. This is somewhat alarming since the increase costs of laparoscopy may be offset if complications such as an expensive ventral incisional hernia repair are decreased in incidence. In the total laparoscopic method the rate of herniation through a trocar site is less than 1% which is a major decline. We continue to use the total laparoscopic Roux-en-Y gastric bypass as our method of choice in morbidly obese patients who weigh under 400 lbs. (Fig. 9.3). The cosmetic advantages are clear but the decrease in incisional hernia and wound infection is a distinct advantage over open surgery and hand assisted surgery (Fig. 9.4). Hand-assisted surgery has a place in the armamentarium of an experienced laparoscopic bariatric surgeon and may serve as a bridge for the novice laparoscopic surgeon. The future of this surgery is to ensure the safety and reliability as has been done with open GBP. Laparoscopic equipment needs to be specially created for this

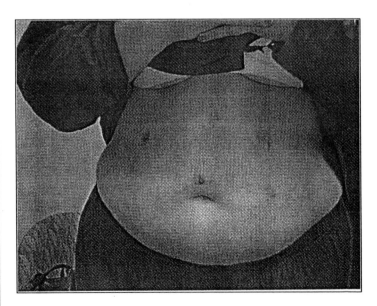

Figure 9.3. Three month postoperative scars after laparoscopic GBP that utilized five trocar sites.

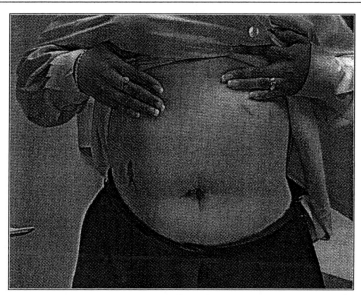

Figure 9.4, lower right. Three month postoperative scars after laparoscopic assisted GBP that utilized three trocar sites and periumbilical Pnuemo Sleeve site.

9

operation and other minimally invasive surgery in the morbid obese patient. The ability to place a hand in the abdomen during pneumoperitoneum is clearly an adjunct to safety and security while maintaining incision length to a minimum.

Selected Readings

1. Whittgrove AC, Clark GW, Schubert KR. Laparoscopic gastric bypass, roux en-Y: Technique and results in 75 patients with 3-30 months follow-up. Obesity Surg 1996;6:500-504.
2. Naithoh T, Gagner M. Laparoscopically assisted gastric bypass surgery using Dexterity Pneumo Sleeve. Surg Endosc 1997;11:830-833.
3. Schweitzer MA, Broderick TJ, Demaria EJ, et al. Laparoscopic-assisted roux-en-Y gastric bypass. J Laproencosc Adv Surg Tech A 1999; 9:449-53.
4. Schweitzer MA, DeMaria EJ, Sugerman HJ. Laparoscopic roux-en-Y gastric bypass. Surgical Rounds 2000; 10:371-380
5. Sundborn M, Gustavsson S. Hand-assisted laparoscopic rous-en-Y gastric bypass: Aspects of surgical technique and early results. Obesity Surgery 2000; 10:420-427.
6. Schauer PR, Ikramuddin S, Gourash W, et al. Outcomes after laparoscopic roux-en-Y gastric bypass for morbid obesity. Ann Surg 2000 Oct;232(4):515-29.
7. Watson DI, Game PA.: Hand-assisted laparoscopic vertical banded gastroplasty. Surg Endosc 1997; 11:1218-1220.
8. Schirmer B. The severly obese patient: Neglected again. J Lap & Adv Surg Tech 1998; 8:1.

CHAPTER 10

Laparoscopic Roux-en-Y Gastric Bypass: Detailed Technical Issues

Eric J. DeMaria

Introduction

Laparoscopic Roux-en-Y gastric bypass is one of the most difficult and challenging laparoscopic procedures routinely performed today. It requires advanced skills and knowledge of both the fields of bariatric surgery and laparoscopy, familiarity with both short and long term follow-up of patients, as well as early postoperative technical and delayed long-term nutritional and metabolic complications. Those issues will be discussed in other chapters. The topic of this chapter is to provide a detailed technical step by step description of the procedure.

Abdominal Access and Creating Pneumoperitoneum

Abdominal access for laparoscopy is a critical issue in the morbidly obese patient. Techniques for abdominal access are wide and varied; however, the most important issue is the careful planning of trocar placement so that the surgeon has the best possible view and can use his surgical instruments optimally to accomplish their objectives. While this principle is very important in all laparoscopic procedures, there is no field in which this is more true than bariatric surgery. Currently we perform laparoscopic Roux-Y gastric bypass (LRYGB) using a 6 access port technique (Fig. 10.1, see all figures at end of chapter). There are critical factors involved in the decision to place each trocar at a given location.

The initial trocar is placed in the left subcostal position. A Veress needle is inserted through an incision large enough to accommodate a 12 mm trocar, just beneath the costal margin. The trocar ideally should be placed in the anterior-axillary line or even a bit more laterally. It may become extremely awkward for the surgical assistant to utilize this trocar if it is placed too medially. After insufflating approximately 4 L of CO_2 gas to create an appropriate pneumoperitoneum, the abdomen is entered under direct vision using one of the commercially available access devices through which the laparoscope is inserted (Visiport®, Tyco/US Surgical, Norwalk, CT). This is particularly useful if the patient has had previous abdominal surgery to avoid bowel injury. The pre-insufflation technique through the Veress needle is important because it creates an intraperitoneal buffer of CO_2 which allows for easy recognition of the appropriate space. Prior to using this technique, we did inadvertently enter the colon during insertion on one occasion, mandating open surgical repair.

Laparoscopic Bariatric Surgery: Techniques and Outcomes, edited by Eric J. DeMaria, Rifat Latifi and Harvey J. Sugerman. ©2002 Landes Bioscience.

This initial 12 mm port is then utilized to introduce a 45° angled viewing laparoscope for placement of the other trocars under direct vision. A 45° scope is very important for successful completion of the procedure, and we have found it impossible to do the procedure with a 0° scope and difficult with a 30° scope. Because high quality images are essential for obtaining the level of technical information that is required to do the procedure,a high quality three-chip camera system is mandatory. We have also found that approaching the larger patient laparoscopically is aided by having a "bariatric-length" laparoscope. We have used several prototypes of this type of scope of approximately 55 cm in length made for our program. (Stryker Endoscopy, Santa Clara, CA). While the length of the scope is very important, one cannot sacrifice on illumination detail, which may occur with a longer scope.

Placement of the supraumbilical trocar for the laparoscope is usually done four finger breadths above the umbilicus and slightly to the left of midline. One should not go so far laterally as to risk injury to the epigastric vessels. This trocar is placed under direct vision after insufflating the abdomen and inserting a 12 mm trocar in the left upper quadrant. This allows one to position the supraumbilical trocar below the anterior abdominal wall attachment of the falciform ligament of the liver. If the trocar comes through the falciform ligament, it is often a constant source of aggravation during the procedure as the fatty tissue may fog the camera repeatedly. This is particularly true when the scope must be pulled back until it is just barely protruding from the trocar during the small bowel portion of the procedure.

Placement of the port site for liver retraction depends on the type of liver retractor used. We prefer a subxiphoid location and utilize a Nathanson liver retraction device which resembles a metal hook inserted through the trocar puncture site after the trocar is removed. Anchoring this device to a rigid fixation arm (Automated Medical Products Corp, Edison, NJ, Iron Intern®) allows for the liver retraction to be accomplished without an assistant holding the retractor, thus keeping it out of the way of the operating surgeon who stands on the patient's right side. There are a wide variety of appropriate liver retractors on the market and most surgeons utilize a right lateral trocar placement for some type of paddle-style liver retractor. These devices can also be anchored with some type of rigid arm system to obviate the need for an assistant to hold the retractor, which may be unreliable during a long surgical procedure.

One additional 5 mm trocar is placed in the left abdomen to allow the surgical assistant to work in a two-handed manner. It is important to assess the amount of abdominal wall distension that has occurred during insufflation in order to determine optimal placement of trocars. If a great deal of abdominal distension has occurred during insufflation, this may cause the subcostal position of the first trocar to descend lower on the abdominal wall than anticipated. Occasionally an 8 to 10 cm gap is created between this trocar and the left costal margin after insufflation. In these circumstances, we insert a 5 mm trocar superiorly at the level of the costal margin on the left side for the assistant's right hand. The assistant will find that this location provides optimal access to the proximal stomach during the initial dissection and gastrojejunal anastomosis. When the abdominal wall is not particularly compliant and less distension occurs, the left-sided initial 12 mm trocar may be within a few centimeters of the left costal margin after insufflation. In that circumstance, we utilize the initial trocar for the assistant's right hand instrument and insert a 5 mm trocar inferior to this location. This trocar should also remain in the lateral abdomen and may be placed slightly medial to the more lateral first trocar in order to offset the assistant's two instruments to some degree and may avoid "sword-fighting" of instruments in working on the proximal stomach.

Locating the "Critical" Trocar for the Surgeon's Right Hand

The most critical trocar for accomplishment of the procedure is the 12 mm trocar for the surgeon's dominant right hand. The surgeon, standing on the patient's right side, must evaluate several factors in positioning this trocar. First, we recommend taking the falciform ligament attachment to the anterior abdominal peritoneum off with ultrasonic dissection during a surgeon's early experience with the procedure. This step will prevent the falciform ligament attachments from forcing placement of the 12 mm trocar for the surgeon's right hand too low and/or lateral. This is critical as the surgeon needs to be close to the midline with the right hand in order to reach the proximal stomach with the length of most available laparoscopic instruments. Specifically, the linear stapling devices available often require insertion of the trocar to the hilt and pressure against the hub of the trocar with the stapling instrument in order to completely transect the stomach at the angle of His. Therefore, the optimal position of this 12 mm trocar for the surgeon's right hand is just off the midline approximately one-third of the distance between the supraumbilical trocar site and the xiphoid process. However, this positioning may need to be moved cephalad in extremely tall patients or in super-obese patients. Finally, the surgeon's left hand 5 mm trocar is placed at the right costal margin in the midclavicular line. A modification for a left hand dominant surgeon is to place a 12 mm trocar for each hand to allow maximum flexibility for the stapler and suturing device (EndoStitch®). This location may need to be moved inferiorly if the patient has an enlarged liver, which is quite common often due to fatty infiltration. It is important to avoid placing a surgeon's left and right hand trocars too close together. There should be at least one hand's breadth between the trocars which can sometimes be accomplished by placing the left hand 5 mm trocar more laterally on the patient's abdomen when the patient is of short stature.

Each trocar inserted for the procedure should have some type of abdominal wall fixation device attached. We then suture all the trocars to the abdominal skin utilizing 2.0 nylon suture. One must avoid dislodgement of trocars during the procedure at all costs as this will create CO_2 leaks in the abdomen leading to excessive gas insufflation and scope fogging. Furthermore, as both surgeon and assistant are utilizing two hands to operate throughout the procedure (Fig. 10.2), it is extremely disruptive to replace a dislodged trocar.

Dissection

The surgical dissection for the perigastric portion of the procedure should only begin after all trocars are positioned and the left lateral segment of the liver is retracted against the diaphragm. The initial dissection (Fig. 10.3) is done through the gastrohepatic ligament to expose the caudate lobe and identify the lesser sac. The stomach is lifted anteriorly and laterally to separate the adhesions between the pancreas and the lesser omentum along the lesser curvature of the stomach. This is done in order to enter the free space of the lesser sac and visualize the posterior gastric wall. It is usually easier taking a more inferior trajectory. Once this is accomplished, we divide the lesser omentum mesentery in order to skeletonize a portion of the gastric wall for a subsequent transection of the stomach. Ideally, we begin transection of the stomach below the second branch of the gastric vein seen anteriorly on the lesser curvature. One can either transect the mesentery on the lesser curvature of the stomach using the ultrasonic scalpel or, often better using a vascular stapler load (2.0 mm staple height) of the linear stapler which provides hemostatic transection (Fig. 10.4). The vagus nerve branches along the stomach are also transected, but we do not believe this has any impact on outcome.

Gastric Transection

Once the lesser curvature of the stomach is adequately cleared, we begin pouch creation by transecting the stomach (Fig. 10.5 and 10.6) using a 60 mm cartridge length of 3.5 mm stapler with the linear cutting stapling device (Endo-GIA II, Tyco/ US Surgical, Norwalk, CT). The aim is to complete the transection at the level of the angle of His with three sequential firings. Prior to the third sequential firing, the posterior portion of the diaphragmatic junction just lateral to the left crus and above the short gastric vessels is dissected in order to clear a path for the stapler to pass (Fig. 10.7). Dissection of the peritoneal junction with the diaphragm anteriorly at the angle of His is also helpful to allow visualization of the cartridge exiting beyond the stomach before the stapler is fired, as complete and hemostatic transection of the stomach must be confirmed (Fig. 10.8 and 10.9). A somewhat tubular pouch is created, based primarily on the lesser curvature by completing the transection right at the angle of His, and one must visualize the completely transected stomach laterally. The goal is to leave a small proximal gastric pouch but at the same time to leave enough gastric pouch tissue in order to create an adequate and secure anastomosis. We estimate the proximal gastric pouch volume to be in the range of 15 to 20 ml. Oozing from the staple line may need to be controlled with clips or interrupted sutures. Clips may interfere with subsequent stapler firing and should be used cautiously. We rarely oversew the staple line in its entirety since this might compromise available gastric tissue to include in the anastomosis.

Lesser Sac Dissection

The gastrojejunal anastomosis is performed routinely posterior to the staple line of the proximal gastric pouch. Therefore, at this point it is helpful to use the ultrasonic scalpel to clear some of the posterior fatty tissue from the pouch to facilitate subsequent suturing. The excluded distal stomach staple line is then grasped by the assistant with each hand and retracted inferolaterally in order to facilitate exposure of the lesser sac inferiorly. The surgeon provides counter-traction with the left hand grasper in the fat just inferior to the pancreas and uses the ultrasonic scalpel to take down any lesser sac adhesions to facilitate subsequent passage of the Roux limb through this area. We place a white, soft rubber drain of 1 inch diameter in the lesser sac which can be easily visualized during mesocolic dissection to mark the path for the Roux limb (Fig. 10.10). Recently, we have begun dissecting the mesocolon from above through the lesser sac in order to place this drain in a location that will facilitate its later identification in the space. To accomplish this, the surgeon's left hand holds the fat in the area of the inferior border of the pancreas. The assistant must maintain one instrument retracting the excluded stomach inferiorly and place the other grasper across from the surgeon's left hand grasper in order to elevate the fatty tissue below the pancreas. The ultrasonic scalpel is used to transect this fatty tissue and overlying peritoneum. Care must be taken to visualize the duodenum in its medial position and prevent it from injury. The dissection is continued unless significant vessels are identified. We open a 2 to 3 cm space with this maneuver and yet rarely transect the mesocolon completely. Some surgeons do completely transect the mesocolon from above with this approach and perform the jejuno-jejunostomy by pulling the bowel up into the lesser sac. We have attempted this on several occasions but abandoned due to difficulties in reducing the anastomosis back through the mesocolon.

Mesocolic Dissection

The end of the rubber drain is placed into the dissected lesser sac where it will be found from below through dissection of the transverse mesocolon. The proximal end of the drain may be sutured to the excluded stomach to prevent its dislodgement. However, this may be unnecessary if a long length of drain tubing is used. We then place the OR table flat and both surgeon and assistant grasp the omentum and use a hand-over-hand movement to retract the omentum into the upper abdomen. The goal of the initial exposure is to identify the mesentery of the transverse colon. This is often only possible by identifying a shining smooth fatty surface, as the omentum and appendices epiploacae of the colon interfere with visualization but usually have a characteristic rough and bumpy surface (Fig. 10.11). This exposure mandates "4 hands", as all 4 graspers must utilized in a coordinated manner.

Once the mesocolon is grasped by one instrument, the other instruments are used to increase the exposure of this location of the mesocolon. Often the assistant's right hand is the most important in this exposure because it is best positioned to hold the transverse colon and omentum cephalad. The ultimate goal for the surgeon's left instrument and the assistant's left instrument is to grasp the mesocolon 1-2 cm lateral and superior to the ligament of Trietz. Occasionally, there is a dimpling of the fat in this area which signals a good location to begin the dissection. Alternatively, an attempt is made to either visualize the rubber drain through the mesocolic fat or to identify discoloration due to blood from the lesser sac dissection which we have termed a "strawberry patch" (Fig. 10.12). If a discoloration is present, the dissection is begun by transecting the peritoneum of the mesocolon over this location. The drain is often found just inside the mesocolon underneath this discoloration.

Opening the mesocolon and finding the lesser sac can be the most difficult step in the laparoscopic gastric bypass procedure. It is certainly the most unpredictable step in our experience as sometimes it goes flawlessly and other times it requires a fair amount of time. To facilitate performance of this step, a repertoire of interventions can be used depending on the circumstances.

The easiest maneuver in our experience is when opening of the transverse mesocolon reveals the rubber drain placed previously. This can be grasped and pulled through the opening approximately one inch, where it is left to mark the mesocolic defect for later passage of the Roux limb. It is important to think in terms of triangulation of the mesocolic opening by three instruments while the fourth moves into the space to identify and grasp the drain tube. The assistant's right hand is usually placed on the anterior-most aspect opening of the mesocolic opening while the surgeon's left hand

and assistant's left hand are placed on the medial and lateral edges of the defect. The key is that only one grasper moves at a time so that the exposure is not lost when two instruments attempt to move simultaneously. Sometimes, the dissection is too posterior, and no free space is identified. Occasionally, one may see the body/tail of the pancreas clearly via this dissection and in those situations one must orient the dissection more anteriorly. Sometimes fairly large openings are required before one can ascertain the correct location.

In the past, we entered the lesser sac from below through the mesocolon and attempted to identify the posterior stomach wall. This proved to be often difficult due to the amount of fatty tissue overlying the stomach. However, if one can identify the posterior stomach wall, it can be grasped by either the surgeon or assistant and retracted inferiorly through the mesocolic defect and then superiorly in order to hold the exposure for additional maneuvering. If the rubber drain has been displaced by these maneuvers, or is not easily visualized but one can identify the posterior stomach, a roticulating grasper can be placed through one of the inferior port sites (i.e., the surgeon's right hand port or the assistant's left hand port) and advance this instrument into the lesser sac behind the stomach (Fig. 10.13). The jaws of this grasper located superiorly near the gastric pouch are used to grasp the drain (Fig. 10.14) and retract it back through the mesocolic defect (Fig. 10.15). Some surgeons do not place anything into this lesser sac location until they have created the Roux limb. This is certainly an acceptable technique but, in our experience, it requires an extra step because the mesocolic area has to be carefully exposed and anatomic landmarks identified twice instead of once.

On two occasions, we have had to abort a posterior retrocolic, retrogastric position for the Roux limb. Each of these patients had very fatty omentum with previous surgery and adhesions of the omentum down in the pelvis and to the small intestine. In this situation, others have described bringing the Roux limb antecolic/antegastric. In order to decrease tension on the Roux, the omentum may be split in a craniocaudad direction and the Roux brought between the cut halves. Alternatively, Champion has described creating an omental window at its junction with the transverse colon. He then brings the Roux limb up through this opening over the colon and stomach to the pouch. In our two cases, we transected the omentum completely and were able to pass the Roux limb to reach the pouch. However, this approach clearly requires more division of the small bowel mesentery to accomplish the greater distance which the Roux limb must travel. It is not clear if these antecolic approaches will be adopted more routinely by others over time, but our experiences have led us to develop a clear preference for the posterior path for the Roux limb.

A final technique worth mentioning is to pass the roticulating grasper from a superior position behind the stomach and to use this is a guide to identify the appropriate location to open in the mesentery of the transverse colon. One difficulty with this approach is that the instrument must push the mesentery in a caudal direction while the surgeon and assistant are trying to lift it anteriorly and superiorly. However, if the tip of the grasper can be identified pressing on the mesocolon, this location can be opened with the harmonic scalpel and facilitate passage of the limb or a Penrose drain.

Creation of the Roux Limb

Once we have marked the location for the mesocolic tunnel, we proceed to perform the small bowel reconstruction for the procedure. The ligament of Trietz is identified with the assistant holding the mesocolon in a cephalad direction in order to facilitate the surgeon's ability to identify this location. A grasper which has marked centimeter increments on it is used to measure 30 cm distal to the ligament of Trietz where the small bowel is transected. The jejunum is kept in the orientation of the letter "C" to avoid disorientation of the limbs (Fig. 10.16). The transection is accomplished with a linear stapler using a 60 mm cartridge of 2.5 mm stapler (Endo-GIA II, Tyco/US Surgical, Norwalk, CT). It is important to make sure that this transection is accomplished perpendicular to the small bowel so that no mesentery is undermined during the transection (Fig. 10.17). This length cartridge will also often divide several centimeters of the mesentery. In our early experience, we felt it necessary to subsequently divide the mesentery further after small bowel transection using a vascular cartridge load of 45 mm length using 2.0 mm staples (Fig. 10.18). With the retrocolic, retrogastric Roux limb passage, this may not be necessary in most cases. We occasionally divide the mesentery with the ultrasonic scalpel until we have approximately 2 to 3 cm transected from the mesenteric edge of the small bowel. We have found this usually to be more than adequate to bring the Roux limb up to the pouch without tension since it is a fairly short distance.

In a surgeon's early experience, it is appropriate to place a marking suture on the afferent limb of the small bowel, i.e., the biliopancreatic limb, but it becomes unnecessary with experience, as the assistant holds this limb of the bowel with the right hand grasper until the anastomosis is begun. The biliopancreatic (afferent) limb is placed on the left side of the viewing screen and we maintain the distal side to the right of the television screen. The distal bowel is pulled toward the operating surgeon with a hand-over-hand technique measuring 50 cm for the Roux limb length. The cut end of this Roux limb should be grasped and pulled toward the operating surgeon and rotated inferiorly in order to maintain the appropriate orientation. A 50 cm Roux limb length is measured for standard proximal gastric bypass procedures (Fig. 10.19).

Jejuno-Jejunostomy

Once the 50 cm mark is identified, a holding suture is placed in the antimesenteric portion of the small bowel and we suture it to the afferent staple line at its antimesenteric end (Fig. 10.20). We use an automatic suturing device to accomplish this (Endostitch, Tyco/US Surgical, Norwalk, CT). A second suture is then placed approximately 2 cm lateral to the first suture to hold the two pieces of bowel together. This suture is held by the assistant's right hand grasper while the surgeon holds the initial suture with his left hand grasper. The surgeon then introduces the ultrasonic scalpel via the right hand port and carefully examines the angle at which this instrument approaches the bowel. The assistant's holding suture is kept in a fixed location and the small bowel rotated left or right by moving the surgeon's left hand grasper either to the left or right. It is particularly important to pay attention to the anterior/posterior orientation of the ultrasonic dissector as many times this anterior/posterior trajectory is the most difficult to replicate in cannulation of the bowel openings with the Endo-GIA stapler in the next step of the procedure.

One centimeter openings are created using ultrasonic dissection in the antimesenteric border of the bowel for insertion of the stapler between the two holding sutures (Fig. 10.21). The bowel wall is grasped and the dissector used to create a full thickness injury which then can be spread in such a manner as to enter the bowel lumen. The dissector provides for good hemostasis in almost every case. The ultrasonic dissector is removed and a linear stapler with a 60 mm cartridge of 2.5 mm staples inserted. We prefer to cannulate these openings with what we call the "pop and drop" technique (Fig. 10.22). The closed stapler jaws is advanced just into each opening simultaneously. "Popping" open the lever on the heel of the stapler will then cause the jaws to "pop" open, retracting the enterotomy sites simultaneously so that the stapler can be easily advanced. It is important to maintain back tension toward the surgeon with the left hand grasper attached to the holding suture. If the stapler will not advance easily into the lumen, this can be facilitated by lifting the stapler anteriorly towards the abdominal wall to prevent it from becoming entrapped posteriorly within the bowel lumen. If the staple jaws can be visualized within the bowel lumen anteriorly then it is often easier to advance the stapler. The assistant's holding suture can also be moved toward the surgeon during this process in order to facilitate pulling the bowel over the stapling device. Once the stapler is clearly identified to be intraluminal from several angled views, the stapler is closed and fired. The intraluminal anastomosis should be briefly inspected but hemostasis is usually excellent. A third Endo stitch suture of 2.0 Surgidac is then placed in the midportion between the previously placed sutures and all three are held anteriorly in such a way that they can be amputated with a firing of the 60 mm cartridge of 2.5 mm staples to close the small bowel enterotomy (Fig. 10.23). To accomplish this, the stapler is introduced into the abdominal cavity and the trajectory and approximation of the stapler is examined and ensured by retraction on the stay sutures. The surgeon is responsible only for the medial holding suture while the assistant's left hand instrument is placed on the middle suture and the right hand instrument placed on the lateral suture. If the GIA stapler takes a sharp anterior/posterior trajectory, it may be necessary to simply lift the surgeon's suture anteriorly and let the other sutures hang free until the last possible moment in positioning the GIA. One must be very careful to avoid amputating too much during this step since this may obstruct the anastomosis. It also is important to note that the knife blade which cuts between the staple lines is not at the anterior-most margin of the stapler, but is rather down the middle of the cartridge where it is marked by a slot on the side of the staple cartridge device. The goal is to have the stay suture knots just visible within the jaws of the instrument. To ensure this the medial location of the stapler for closure is positioned first followed by the lateral tip. With the U.S. Surgical stapler, tissue gap control allows the surgeon to reposition tissue at the farthest extent (tip) of the stapler's jaw even though the device is closed. This allows us to manipulate the tissue and decrease the amount amputated laterally without losing the important proximal positioning that was obtained. Ideally, a very small rim of tissue is amputated to just include the previously tied three sutures knots. This tissue specimen is then removed through a 12 mm trocar and discarded.

The anastomosis and enterotomy closure sites are inspected with particular attention to the medial corner where the two staple lines intersect. We have occasionally found small defects in the staple line at this area even though the enterotomy closure appears to have been carried out without difficulty. One must inspect the bowel on either side and expose this area of staple line intersection to make sure that there is no hole. This is also the location for placement of a suture to avoid kinking of the bowel wherein the Roux limb can become twisted or adhesed to the afferent limb staple line creating obstruction. Similarly, traction on this suture provides retraction for closure of the mesentery of the jejuno-jejunostomy anastomosis (Fig. 10.24). The surgical assistant retracts this suture anteriorly and the surgeon places a 9-inch Endostitch suture at the base of the mesentery at the apex of the mesenteric opening and this suture is run anteriorly and tied to the holding suture. In addition, we reinforce the stapled anastomosis with one additional interrupted suture at the "toe" of the GIA anastomosis between the afferent and efferent limbs. It is also critical to avoid crimping the mesentery of the Roux limb with the mesenteric closure sutures in a way that the Roux limb mesentery is foreshortened and would be unable to reach to the proximal gastric pouch without tension. It is once again important to inspect the alignment of the limbs and the surgeon should confirm that the Roux limb proceeds toward him (on the patient's right side) with the afferent and efferent limbs directed away to his left and right, respectively. Any variance from this orientation should lead the surgeon to more detailed examination of the anatomical situation. On several occasions, we have identified unsuspected errors in creation of the anastomosis when this alignment was not confirmed. For example, no bowel should come through the mesenteric defect. This always signifies an error in creation of the anastomosis and will likely lead to postoperative obstruction. It is much better to take down the anastomosis and redo it than to deal with a postoperative obstruction or anastomotic leak.

Bringing the Roux Limb through the Retrocolic Tunnel

The Roux limb is examined from the jejuno-jejunostomy back to its cut end. A suture is placed between the end of the Roux limb approximately 2 to 3 cm from the staple line to the drain which was previously placed through the mesenteric defect. By placing a suture on this location it is easier to identify the orientation of the limb as it comes up through the retrogastric tunnel. The limb is placed with the orientation of the small bowel towards the surgeon and the cut end of the mesentery towards the assistant laterally and carefully guided up through the mesocolon (Fig. 10.25). The patient is then placed back into steep reverse Trendelenburg position on the table and the drain is visualized behind the stomach, grasped and delivered with the attached Roux limb using a hand-over-hand technique by the surgeon and assistant with combined effort. Oftentimes, the assistant must use his left hand to retract the excluded stomach down and enhance visibility as the limb comes through the mesocolic opening. Proper orientation of the limb must be confirmed as the limb is delivered from the retrogastric position with the stapled end of the small bowel facing the surgical assistant laterally and the Roux limb towards the surgeon. If this orientation is not confirmed, it is imperative to go back to the mesocolic passage and make sure the limb is not twisted. It is much better to redo this step than to have an obstructed and twisted Roux limb in the retrogastric position where it is not easily visualized.

The Roux limb should be easily delivered without tension and lay next to the gastric pouch without any need to hold it there by the assistant or surgeon. If there is tension on the limb, it suggests a twist in the blind passage behind the stomach and this must be examined and corrected. Alternatively, there may be a band of tissue in the retrogastric passage that is preventing the limb from being delivered without tension. Both these possibilities must be sought out and treated. Ultrasonic dissector can be used to transect any bands of tissue limiting the Roux limb passage, and it should then be delivered to the gastric pouch without tension. The gastrojejunal anastomosis can then be undertaken.

Gastrojejunal Anastomosis

Four techniques have been described for performing the gastrojejunal anastomosis in this procedure:

1. circular stapler with transesophageal passage of the anvil,
2. circular stapler with transgastric insertion of the anvil,
3. linear stapled anastomosis, and
4. hand sewn anastomosis.

Although currently, our technique of choice to create gastro-jejunostomy is the linear stapled anastomosis we will describe the other techniques available to allow the reader to gain understanding of other options. It is helpful to have an available repertoire of techniques to deal with difficult situations that can arise intraoperatively.

Circular Stapler Anastomosis with Transesophageal Passage of the Anvil

The original technique for performing this anastomosis, as described by Wittgrove and Clark, was based upon the concepts of the percutaneous endoscopic gastrostomy (PEG) procedure for transoral placement of a circular stapler's anvil into the proximal gastric pouch via the esophagus (Fig. 10.26). This technique overcame several difficult obstacles by allowing creation of a small gastric pouch without fear that there would be inadequate gastric tissue to create a reproducible small diameter anastomosis, as well as allowed surgeons to avoid laparoscopic suturing, which was perceived in the mid 1990s as an onerous task. After transection of the stomach, the pouch is observed laparoscopically while a surgical assistant passes a flexible gastroscope per os until the scope's light is visualized within the gastric pouch. A snare is advanced through the scope to strike the pouch wall where it is directed to an appropriate location within the pouch for the center of the planned anastomosis. This snare is often aimed at a location posterior to the gastric staple line and ideally planning to avoid overlapping the circular staple line with the line of staples from the gastric transection. The importance of this planning is unclear, however. Electrocautery is then applied directly over the site of the endoscopic snare while the assistant advances it until it pops through the wall of the stomach into the laparoscope's view. The surgeon inserts the needle of the PEG kit transcutaneously into the abdominal cavity under direct laparoscopic visualization followed by the wire which is grasped by the endoscopist's opened snare allowing the scope to be removed and the wire pulled up and out the patient's mouth. The anvil stem is then attached to the looped end of the PEG wire so that it can be pulled down through the mouth and esophagus and through the wall of the gastric pouch. Guiding the anvil through the soft tissues of the oropharynx can be difficult and is facilitated by performing an anterior jaw lift maneuver while the surgeon pulls steadily on the wire from below.

Occasionally it is necessary to use a rigid laryngoscope to visualize the anvil deep in the pharynx and push it forward with a finger or instrument. Deflation of the endotracheal tube balloon may also be helpful in some situations, although in some cases it is difficult to place the anvil successfully via the mouth despite these maneuvers. In our early experience, we used a 96-inch long heavy non-absorbable suture in place of the PEG wire to draw the anvil down the esophagus since we felt opening a PEG kit to be an unnecessary expense. In one case, however, this long suture broke while pulling the anvil leaving it beyond reach in the mid-esophagus. This challenging problem was solved by creating a gastrotomy in the medial pouch and advancing a sterile esophageal dilator retrograde up the esophagus until it contacted the anvil and allowed us to push it back out through the mouth. Several authors have reported non-transmural esophageal tears using the PEG wire technique.

Further application of electrocautery is usually required to allow the anvil to pop through the gastric wall, but care is required to avoid making too large an opening, although one can suture this opening closed around the stem with laparoscopic technique should a large opening occur. The circular anastomosis is created by inserting the stapling device (Ethicon Endosurgery, Cincinnati, OH) via the abdominal wall by removing and dilating the tract of a 12 mm abdominal port site. The stapler is advanced into the Roux limb via an enterotomy incision in the bowel created with the ultrasonic dissector. The two components of the stapler are then mated using laparoscopic instruments to facilitate the proper orientation by fixing the anvil stem in a stable position and advancing the stapler as needed. The two pieces 'snap' together and the stapler can then be closed and fired. We found one of the most challenging aspects of the circular stapled technique to be the safe removal of the stapler after firing. The stapler is opened according to the manufacturer's instructions (which differ according to the product being used), and the stapler is rotated left and right in order to free it from the tissue and then withdrawn. It is very difficult to determine how much back tension on the stapler is safe to apply, i.e., the necessary backward force to dislodge the stapler without tearing the anastomosis. This single step can create much anxiety for the surgeon during the learning curve.

Circular Stapler with Transgastric Insertion of the Anvil

As a direct result of concerns with safe placement of the anvil via the transesophageal route, we (and others) modified the technique by which the anvil is inserted via a gastrotomy either before or after transection of the stomach. The gastrotomy can be created anteriorly in the body of the stomach and the anvil inserted on a long grasper into the proximal gastric lumen where an opening is made with electrocautery to allow the stem to be advanced into position, followed by transection of the stomach, and ultimately stapled closure of the gastrotomy. Alternatively, a catheter can be sutured to the end of the anvil stem and brought through the area of proposed anastomosis. Attaching a suture with a straight needle on one end to the anvil stem at its other end and piercing the gastric wall from an intraluminal to extraluminal direction may be used. Pulling on the suture allows one the delivery of the anvil into the correct position. Finally, the stomach may be transected first and a subsequent gastrotomy made in the pouch for insertion of the anvil attached to a suture, a catheter, or simply placing the stem in position followed by cauterization of this location to push the stem through the gastric wall. We tried each of these techniques and were generally dissatisfied with the size of the gastric pouch which we tended to make larger than an ideal size because of our concerns about either stapling the stomach around the anvil or because we planned to make a gastrotomy for anvil insertion in the gastric pouch.

In addition to this technique being awkward it requires dilatation of the abdominal wall to insert the circular stapler as described above. Furthermore, postoperative pain following this port site dilatation maneuver may prolong hospital stay in some patients following surgery. As reported by others, we also noted higher wound infection rate at this port site skin incision likely resulting from bringing the tissue donuts through this wound when the stapler was removed. Some authors have recommend a mechanical and antibiotic bowel prep before surgery to decrease the risk of wound infection. The concerns of postoperative pain and wound infections do not seem related to the method of anvil placement but rather to the concept of using a circular stapling device to create the anastomosis no matter how the anvil was delivered into the gastric pouch.

Linear Stapled Gastrojejunal Anastomosis

This technique was developed by Champion although, some surgeons have utilized this technique for the gastrojejunal anastomosis during open bypass surgery. In our early experience with this technique, we placed two holding sutures with the Endo-stitch followed by creating 1 cm openings in the Roux and in the gastric pouch with the ultrasonic dissector for the the linear stapler (Endo-GIA II, Tyco/USSurgical, Norwalk, CT) to be inserted approximately 2.5 cm to create a side-to-side gastroenterostomy. The anastomosis was then stented with an appropriately small (10-12 mm, 30 or 32 F) diameter dilator passed via the mouth by the anesthesiologist. In a number of cases, we used a subsequent firing of the linear stapler to close the anterior opening in the anastomosis over the internal stent. Experience with this technique was unsatisfactory for two reasons: 1) the intraoperative leaks during air insufflation via the endoscope were unacceptably high (20%) frequency of cases and 2) sometimes it was difficult to pass the scope through the anastomosis because of narrowing along the long closure staple line. As a result of these concerns, we began to oversew the entire anastomosis to prevent the need for intraoperative leak repair. Currently, we begin by running a non-absorbable (2-0 Surgidac® on the Endo-stitch®, Tyco/USSurgical, Norwalk, CT) 9-inch length suture between the Roux limb and the posterior gastric pouch for a distance of approximately 3 cm to serve as a posterior suture row for a 2-layer anastomosis (Fig. 10.27). The stapler is then used to create an 'inner row' of the anastomosis posteriorly (Figs. 10.28 and 10.29). We have found that a 45 mm length cartridge of 3.5 mm staples provides adequate staple height for the thickness of the tissue and is hemostatic for this inner row. Using a shorter cartridge length, e.g., 30 mm, is not desirable as the stapler itself may not be long enough to reach the appropriate location due to the patient's obesity.

Rather than an esophageal dilator, we routinely guide the flexible gastroscope down the esophagus and across the anastomosis without insufflation of air to avoid bowel distension during the procedure, leaving it in position for subsequent insufflation of the bowel lumen to visualize and test the integrity of the surgical anastomosis. This saves time, as it can be difficult to pass the scope into the gastric pouch from above and immediately find the completed gastroenterostomy due to bloody fluid within the lumen. The endoscopist passes the gastroscope by visualizing the scope's light transilluminating the gastric pouch wall on the laparoscopic monitor (Fig. 10.30) without looking through the endoscope itself. The scope is passed into the Roux limb and well below the site of anastomosis where it is left to serve as a stent. The scope diameter of 11 mm is appropriate for sizing / stenting and allows us to rapidly withdraw the scope after clamping the Roux limb (to prevent bowel distension) without delay.

The open area of the gastrojejunal anastomosis is closed with another running suture of 2-0 Surgidac® using the Endo-stitch (Fig 10.31). We then oversew the entire anterior anastomosis with a final running suture of 2-0 Surgidac® starting from the lateral holding suture (placed during the intial suture layer posteriorly) and run medially to the medial holding suture where it is tied. We use a 12 inch length of suture to accomplish this. We often are able to incorporate the gastric pouch staple line in this final running suture, adding security because the tissue bites are quite adequate with this technique as well as reinforcing the staple line with suture. Using this oversew technique, our intraoperative leak rate during the air insufflation test has decreased to 5% and our postoperative leak rate to less than 2%.

As mentioned above, the integrity of the surgical anastomosis is routinely tested by submerging the anastomosis area under saline solution after placing the atraumatic intestinal clamp across the limb and insufflating air through the gastroscope (Fig. 10.32). The clamp creates an obstruction to allow tense distension of the limb as well as to prevent air insufflation from causing dilatation of the bowel and excluded stomach which interfere with visualization for completion of the procedure. We have not measured the intraluminal insufflation pressures obtained during the gastroscopy but our impression is that good distension is easily accomplished. Although some authors have filled the pouch with methylene blue in order to test for leaks, adequate distension is unlikely to occur without occluding the egress of contrast retrograde into the esophagus. This problem is partially overcome by the ability to continuously insufflate air through the gastroscope as opposed to the fixed-volume used with instillation of methylene blue dye. Following the endoscopic insufflation test the air is suctioned out of the Roux limb with the gastroscope. We remove the intestinal clamp and place a 10 mm flat closed-suction drain posterior to the gastrojejunal anastomosis as the final step of the procedure (Fig. 10.33). The drain is brought out through the right upper quadrant port site and secured. It is left in position as a "monitor" for a postoperative leak and is removed routinely on postoperative day 2 after a normal contrast x-ray study of the anastomosis and the patient is tolerating a pureed diet. The drain is only left in place if there is concern about a leak or if the output is greater than 100 cc per shift, and on those rare occasions the drain is left until these concerns resolve.

Closure of Mesentery

This critical step in completion of the gastric bypass procedure has been omitted by many surgeons, both in open and in laparoscopic surgery. However, any surgeon who has seen a patient suffer the devastating complication of gut infarction from a strangulated internal hernia is unlikely to believe that this closure is unnecessary. We have seen several patients with postoperative abdominal pain attacks following laparoscopic gastric bypass surgery in which repeat laparoscopy revealed a mesenteric defect despite operative closure of the defect at the time of the bypass procedure. Although, it is possible that the incidence of internal hernia might increase following laparoscopic Roux-en-Y gastric bypass due to the paucity of intra-abdominal adhesions with this access technique, making the bowel more mobile and able to herniate more easily,this theory remains unproven.

There are three mesenteric defects which arise as part of the retro-colic retro-gastric Roux-en-Y gastric bypass procedure (Fig. 10.34). Repair of the defect at the jejuno-jejunostomy was previously described in the section dealing with creation of that anastomosis. At the transverse mesocolon, there are two mesenteric defects;

1. the mesocolic defect leading into the lesser sac, and
2. the Petersen 's defect, i.e., the space posterior to the cut mesentery of the roux limb (Fig. 10.35).

These can be repaired together and we currently do so by suturing the Roux limb mesentery to the mesentery of the transverse colon. As a result of our concerns about internal hernias at the mesocolic defect, we have developed a more extensive repair technique to oversew the defects than most surgeons perform. The omentum is retracted cephalad and the operating room table once again placed flat. The transverse mesocolon is grasped and retracted anteriorly by the surgeon and assistant each using both hands to facilitate exposure of the mesenteric defect for passage of the Roux limb.

The closure is accomplished with a 9-inch suture placed laterally between mesentery of the Roux limb and the transverse mesocolon, and we run this suture medially including placing sutures in the bowel tissue itself. The sutures placed in the bowel are locked to avoid a purse-string effect which might cause obstruction of the Roux at the level of this closure. The running suture is terminated at a convenient location after several "bowel to mesocolon" sutures have been placed. The assistant retracts this holding suture laterally and anteriorly to allow visualization of the Petersen's defect. A second 9-inch Surgidac® Endostitch suture is begun at the apex of this defect posteriorly and run anteriorly with bites between the mesenteries until it can be tied to the previous holding suture. The assistant must create and maintain adequate exposure during this process, particularly retracting omentum or other fatty tissues that may obscure visualization. Occasionally this mandates dropping the initial holding suture in order to expose the defect adequately. This is not a problem if a long tail of suture has been left so that it does not become displaced.

Closure of the Fascial Defects

We previously were not routinely closing the fascial defects at each trocar site. We began routinely closing 12 mm trocar fascial defects as a result of having two acute postoperative bowel obstructions due to trocar site hernias. We have not seen a single wound infection in our patients since we adopted the linear stapled technique for the gastrojejunal anastomosis. Thus, our current technique has virtually eliminated wound complications in our bariatric patients and this provides the major measurable outcome benefit for the laparoscopic procedure.

Summary

The laparoscopic Roux-en-Y gastric bypass procedure is technically demanding and should not be underestimated by the novice surgeon. Understanding the technical nuances of the procedure, some of which are detailed in this chapter, may help surgeons accomplish this procedure safely. This description, however, is no substitute for other educational pursuits, including postgraduate courses, hands-on animate laboratory sessions, inanimate skill training sessions, preceptoring sessions, and fellowship training in the area of laparoscopic bariatric surgery. The interested surgeon should seek out each of these training opportunities in order to perform the procedure safely in this challenging patient population.

Suggested Readings

1. Higa K, Boone K, Ho T, Davies O. Laparoscopic Roux-en-Y gastric bypass for morbid obesity. Arch Surg 2000; 135:1029-1034.
2. Nguyen N, Ho H, Palmer L, Wolfe B. A comparison study of laparoscopic versus open gastric bypass for morbid obesity. J Am Coll Surg 2000; 191:149-157.
3. Schauer P, Ikramuddin S, Gourash W. Ramanathan R. et al. Outcomes after laparoscopic gastric bypass for morbid obesity. Ann Surg 2000; 232(4): 515-529.
4. Schweitzer MA, DeMaria EJ, Sugerman HJ. Laparoscopic Roux-en-Y gastric bypass. Surgical Rounds 2000; 371-380.
5. Schweitzer MA, DeMaria EJ, Broderick TJ, Sugerman HJ. Laparoscopic closure of mesenteric defects after Roux-en-Y gastric bypass. J Lapendosc and Adv Surg Techniques 2000; 10(3):173-175.
6. Wittgrove A, Clark G, Tremblay L. Laparoscopic gastric bypass, Roux-en-Y: preliminary report of five cases. Obes Surg 1994; 4:353-357.
7. Wittgrove A. Clark G. Laparoscopic gastric bypass, roux en-y-500 patients: Technique and results, with 3-60 month follow-up. Obesity Surg 2000; 10: 233-239.
8. DeMaria EJ, Sugerman HJ, Kellum JM et al. Results of 281 consecutive total laparoscopic Roux-en-Y gastric bypasses to treat morbid obesity. Ann Surg 2002 (in press).

10

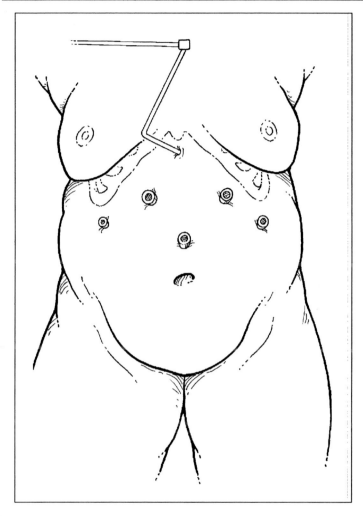

Fig. 10.1. Schematic demonstrating trocar insertion sites for total laparoscopic gastric bypass.

10

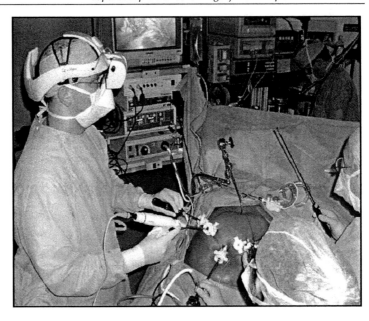

Fig. 10.2. Intraoperative photograph demonstrating ideal positioning of the surgeon's hands during laparoscopic gastric bypass.

10

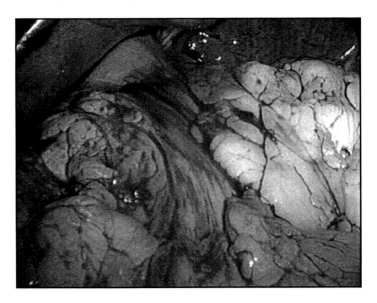

Fig. 10.3. Initial perigastric dissection is through the gastrohepatic ligament to expose the caudate lobe of the liver and the lesser sac.

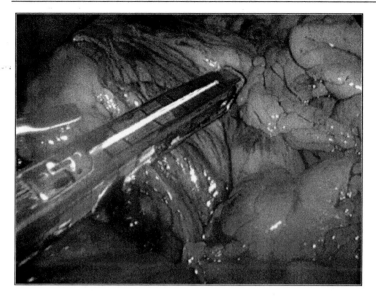

Fig. 10.4. The gastric transection is begun on the lesser curvature aspect of the stomach approximately 3 cm below the gastroesophageal junction using the linear stapler with a 60 mm cartridge of 3.5 mm staples.

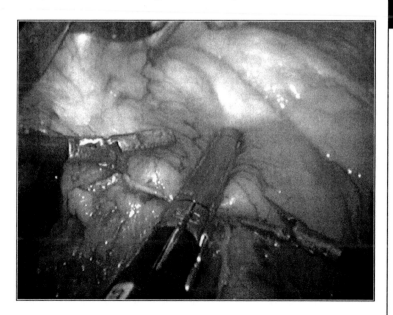

Fig. 10.5. Sequential firings of the stapler are used to divide the stomach.

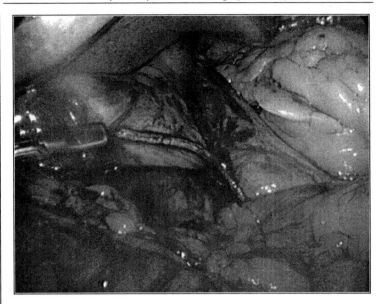

Fig. 10.6. The small size of the proximal gastric pouch is well seen in this intra-operative photograph.

10

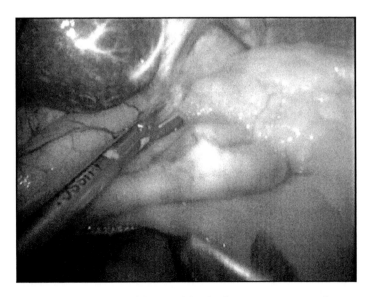

Fig. 10.7A. Dissection at the left crus of the diaphragm anterior (A) and posterior (B) to the stomach at the angle of His is utilized to facilitate passage of the stapler for the final firing to complete the transection.

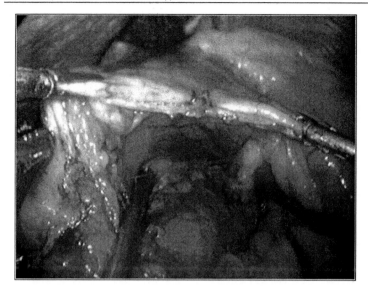

Fig. 10.7B.

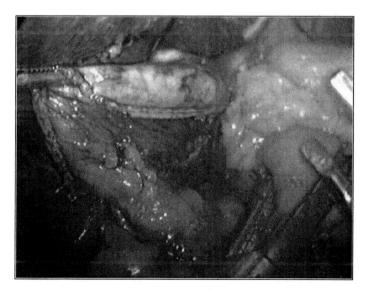

Fig. 10.8. The gastric transection is completed at the angle of His and the surgeon must adequately visualize the complete transection of all gastric tissue.

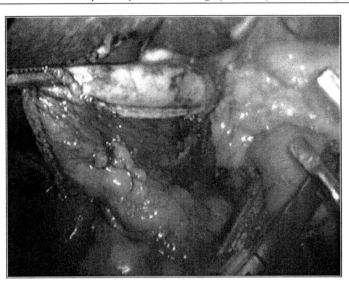

Fig. 10.9. The posterior wall of the proximal gastric pouch is cleared of overlying fatty tissue without devascularizing the pouch, particularly on the medial aspect where the primary blood supply is found.

10

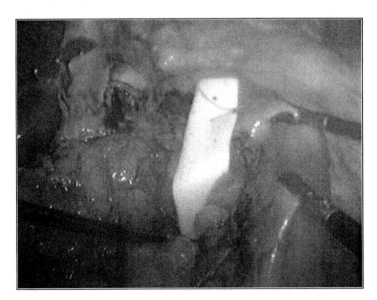

Fig. 10.10. A soft rubber latex-free one inch diameter drain is placed into the lesser sac posterior to the bypasses stomach remnant to facilitate identification of the lesser sac in the subsequent mesocolic dissection.

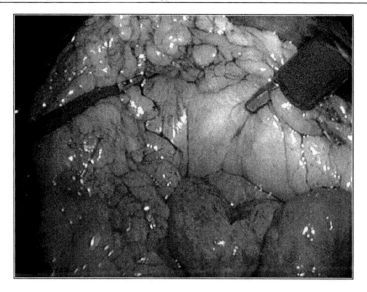

Fig. 10.11. The transverse colon and greater omentum are retracted superiorly to expose the transverse colon mesentery which is grasped and retracted anteriorly and in a cephalad direction.

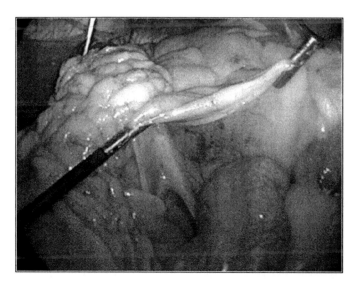

Fig. 10.12. Dissection in the lesser sac behind the bypassed stomach may create this 'strawberry patch' in the transverse mesocolon which provides a location where it is possible to begin the mesocolic dissection, usually 1-2 cm lateral and superior to the ligament of Treitz.

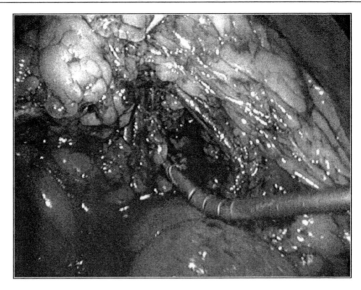

Fig. 10.13. On those occasions where the rubber drain can not be identified in the free space exposed by entering the mesocolon, it is helpful to place a long roticulating instrument from below into this space.

10

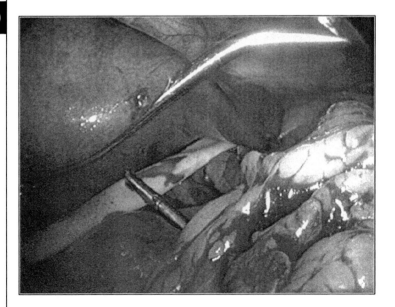

Fig. 10.14. The roticulating instrument can then be utilized to grasp the drain and pull it back through the mesocolon.

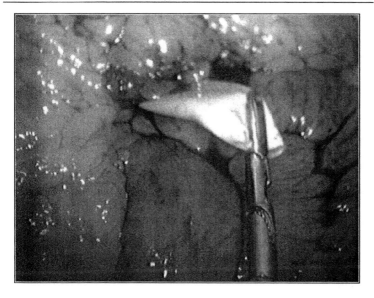

Fig. 10.15. The drain is left in position marking the dissected mesocolic pathway for subsequent passage of the Roux limb

10

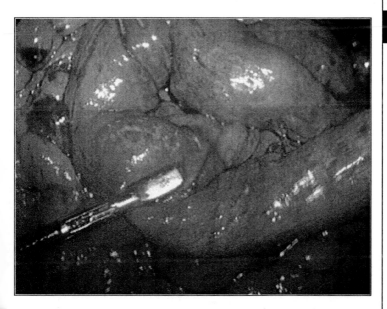

Fig. 10.16. The small bowel is then measured 30 cm distal to the ligament of Treitz where it is transected. The optimal orientation is to keep the small intestinal loop to be divided oriented in the shape of the letter "C".

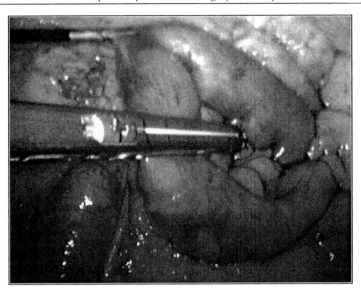

Fig. 10.17. The proximal small intestine is transected perpendicular to the mesenteric border with the 60 mm cartridge of 2.5 mm staples.

10

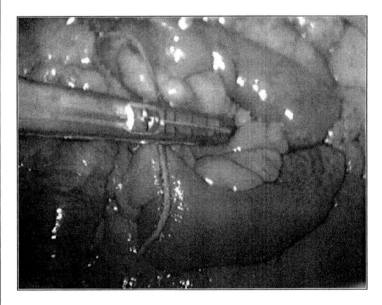

Fig. 10.18. Subsequent transection of the small bowel mesentery may be accomplished with either the ultrasonic dissector or the vascular cartridge 45 mm length of 2.0 mm staples.

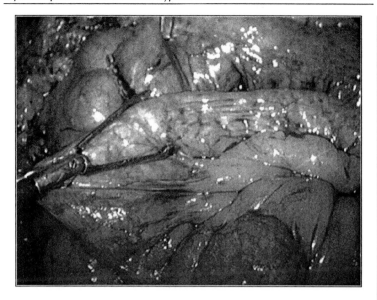

Fig. 10.19. The Roux limb length of 50 cm is then measured using hand-over-hand technique.

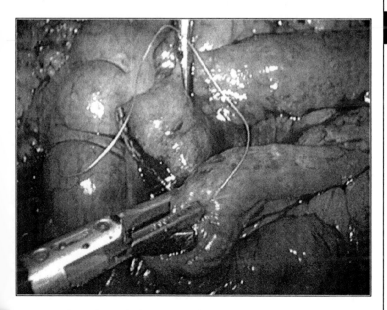

Fig. 10.20. The Endo-Stitch device is used to place a holding suture between the afferent biliopancreatic limb and the distal small bowel. The correct orientation of the three bowel limbs must be preserved.

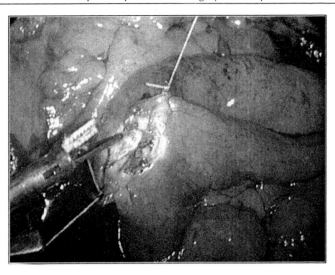

Fig. 10.21. The ultrasonic dissector is utilized to create a 1 cm enterotomy in each small bowel segment one-half way between the holding sutures.

10

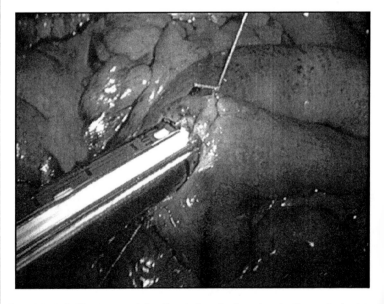

Fig. 10.22A-C. The "pop and drop" technique for inserting the linear stapler into the enterotomy sites. A) the closed stapler is introduced and placed so that each jaw is gently insinuated into the small bowel openings created previously.

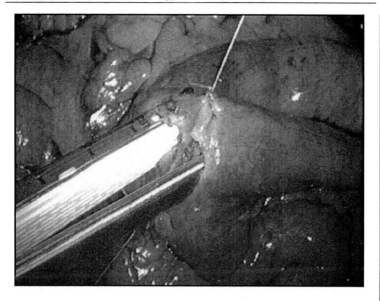

Fig. 10.22B. "Popping" open the stapler allows the jaws to retract each opening simultaneously so that the stapler advances into the bowel lumen by gentle pressure.

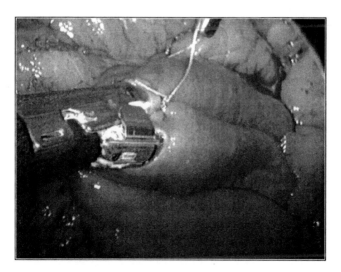

Fig. 10.22C. The stapler is advanced with counter traction applied to the holding suture closest to the surgeon and while being careful to visualize the jaws anteriorly within the bowel lumen.

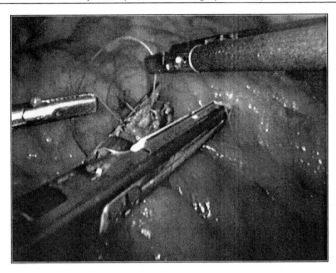

Fig. 10.23A-C. Amputation of the enterotomy used to create the anastomosis is facili-tated by placement of a third holding suture midway between the initial two sutures. A) The stapler is oriented such that it parallels the knots seen on each holding suture in both the anterior-posterior and left-right directions. B) The stapler is closed and in-spected to determine that each knot is visible. C) The stapler is rotated to confirm that adequate tissue is being amputed to ensure closure, but that excessive tissue is not included in the jaws which would narrow the anastomosis.

10

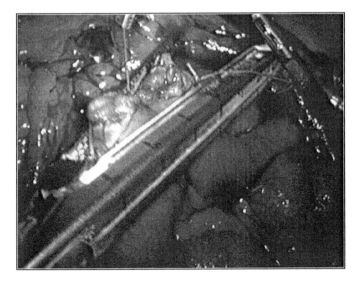

Fig. 10.23B.

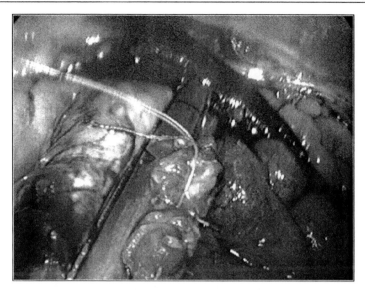

Fig. 10.23C.

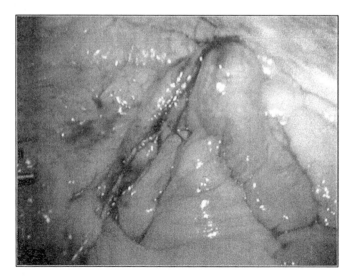

Fig. 10.24. The staple line is inspected to ensure complete closure. A holding suture is then placed between the 2 pieces of bowel to reinforce the staple line. Retraction of this suture anteriorly by the assistant allows visualization of the mesenteric defect. The mesenteric defect is closed with running suture (2-0 Surgidak®, Tyco/USSurgical).

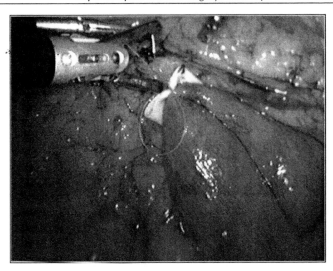

Fig. 10.25. A) The soft rubber drain protruding from the mesocolon is attached to the Roux limb by suture. B) The Roux limb is then advanced into the mesenteric defect without altering its orientation in order to prevent twisting or kinking. C) The drain is retracted anteriorly and cephalad above the bypassed stomach in order to deliver the roux limb into proximity with the proximal gastric pouch without tension. D) Correct orientation of the limb is confirmed once it has passed up through the tunnel.

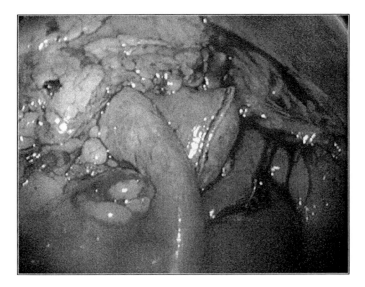

Fig. 10.25B.

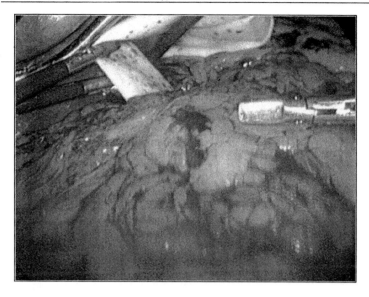

Fig. 10.25C.

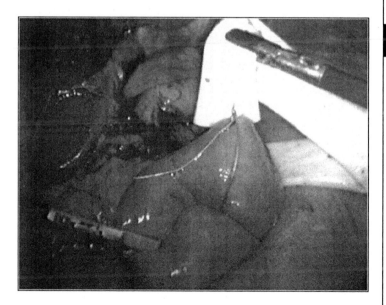

Fig. 10.25D.

10

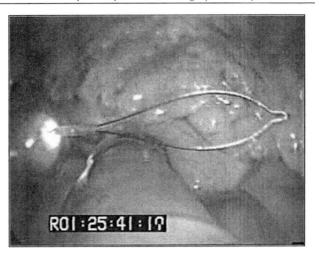

Fig. 10.26A-D. The PEG technique of gastrojejunal anastomosis. A) The flexible endoscope in the gastric pouch is utilized to pass a snare through the pouch wall and into the abdominal cavity.

10

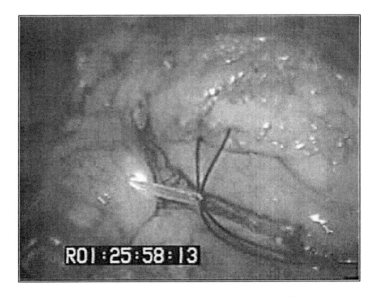

Fig. 10.26B. The PEG technique of gastrojejunal anastomosis. B) The snare is utilized to grasp the wire which is advanced up the esophagus to exit the patient's mouth.

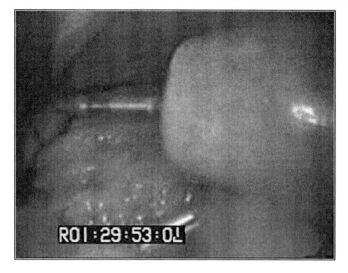

Fig. 10.26C. The PEG technique of gastrojejunal anastomosis. C) The circular stapler is inserted via an enterotomy into the Roux limb and the spike is utilized to penetrate the small bowel through the anti-mesenteric side.

10

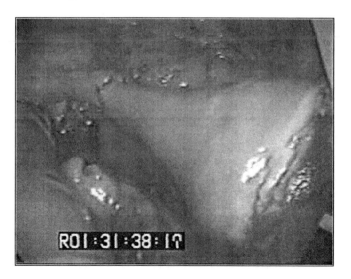

ig. 10.26D. The PEG technique of gastrojejunal anastomosis. D) After firing, the cir-ular stapler is removed with gentle back tension on the device.

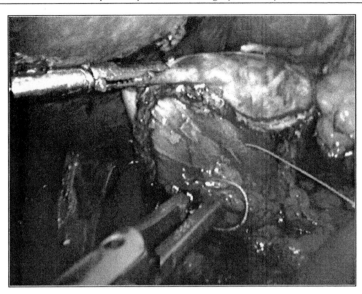

Figs. 10.27A and B. For the linear stapled gastrojejunostomy technique, initially a holding suture is placed between the posterior pouch and the Roux limb medially (A), followed by a running posterior suture layer (B).

10

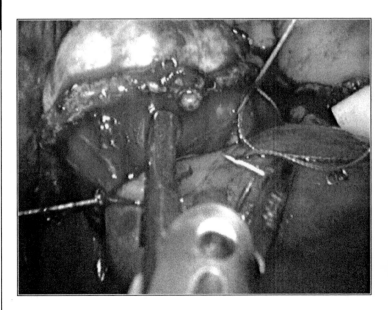

Fig. 10.27B.

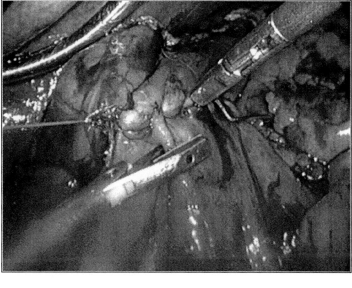

Fig. 10.28A. One centimeter openings are made in the Roux (A) And in the gastric pouch (B) using the ultrasonic dissector .

10

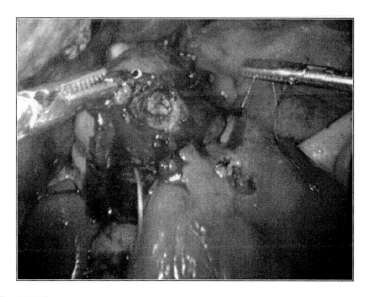

ig. 10.28B.

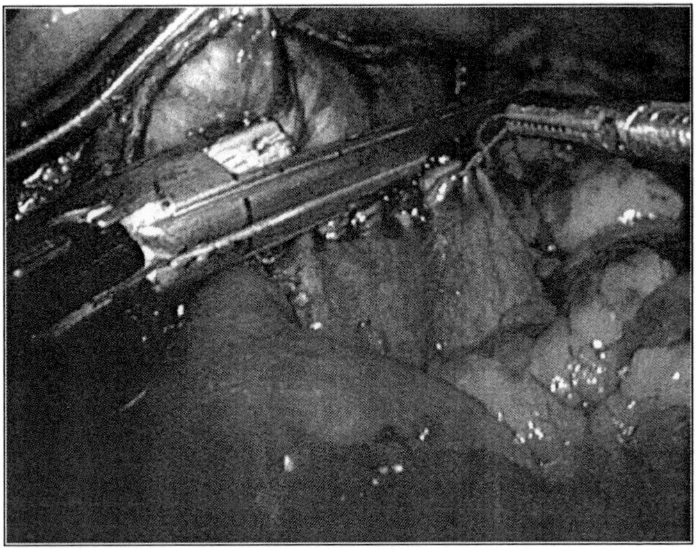

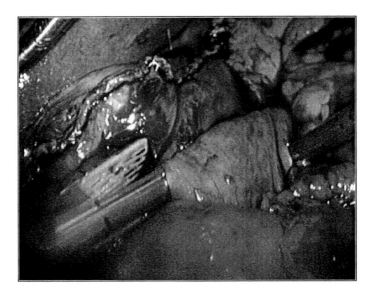

Fig. 10.29A. A 45 mm cartridge of 3.5 mm staples is inserted in the gastrotomy initially (A) in order to size the opening, followed by insertion of both jaws (B) to a distance of 2.5 to 3 cm to create the anastomosis

10

Fig. 10.29B.

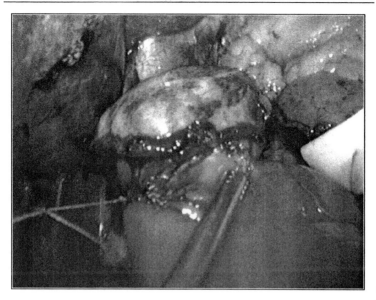

Fig. 10.30. We insert the flexible gastroscope into the pouch and through the anastomosis by looking at the laparoscopic view of the transilluminated scope light in order to avoid insufflation of the bowel.

10

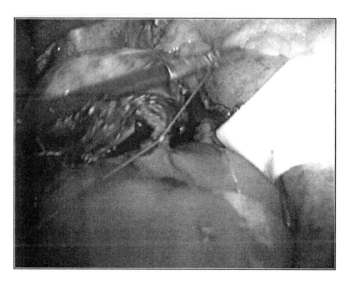

Fig. 10.31. The opening in the anastomosis is then sutured closed with an inner (A) and outer (B) running suture.

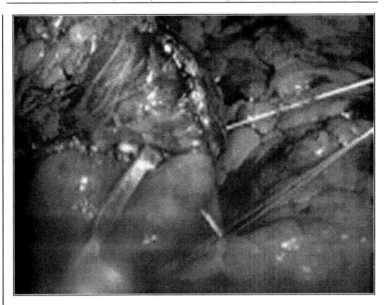

Fig. 10.31B.

10

Fig. 10.32. The Roux limb is occluded with an atraumatic bowel clamp to allow air insufflation via the gastroscope in order to test for air leaks when the anastomosis is submerged under saline.

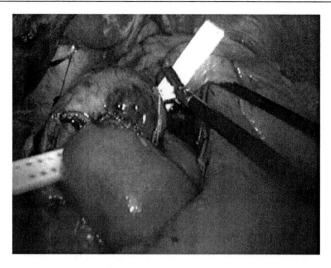

Fig. 10.33. A closed suction drain is placed posterior to the gastrojejunal anastomosis and left in place until a postoperative contrast study is performed on the first postop day and the patient tolerates a pureed diet.

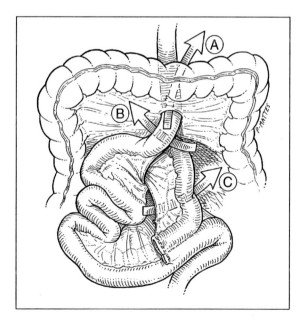

10

Fig. 10.34. There are three mesenteric defects that must be sutured closed in the Roux-en-Y gastric bypass procedure.

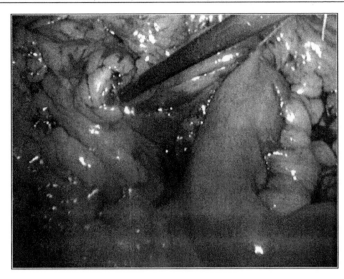

Fig. 10.35A-C. The Petersen 's defect posterior to the Roux limb mesentery is best visualized by retraction of the Roux and the mesocolon anteriorly and looking from right to left (A, above) Closure of the mesocolic defect is undertaken with an initial running suture placed laterally between the Roux mesentery and the mesocolon followed by a running suture placed from medial to lateral to complete the closure (B and C).

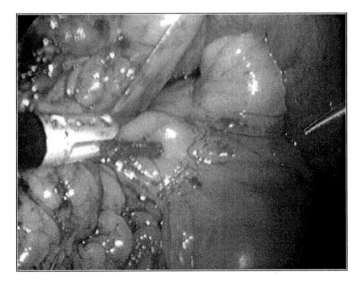

Fig. 10.35B.

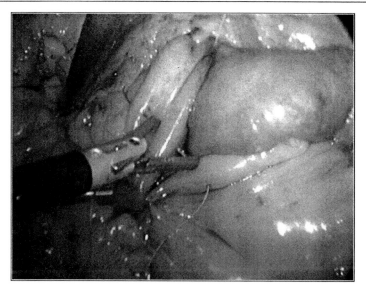

Fig. 10.35C.

10

CHAPTER 11

Laparoscopic Gastric Bypass: Clinical Outcomes

Shanu N. Kothari and Eric J. DeMaria

Introduction

In 1993, Wittgrove and Clark performed the first laparoscopic gastric bypass. This was soon followed by a published report in 1994 of their initial five cases outlining the technical feasibility of the operation.[1] Since that time variations in operative technique have been described, the majority of which differ in the type of gastrojejunal anastomosis created. Regardless of the laparoscopic technique used during the procedure, the questions that must be answered pertain to outcomes in comparison to open gastric bypass. In this chapter, the world's literature will be reviewed pertaining to outcomes following laparoscopic gastric bypass. As of this printing, the following five papers make up the published experience regarding laparoscopic gastric bypass outcomes and are presented in the order they were published.

Literature Review

In 1998, Lonroth and Dalenback published their experience with a variety of laparoscopic bariatric procedures.[2] Included in their experiences were 29 patients who underwent laparoscopic gastric bypass. Their initial ten were, in fact, an omega loop gastroanastomosis followed by 19 true Roux-en-Y gastric bypasses. Their pouch size is measured using a ruler from the angle of His to the lesser curve, preferably 4 cm in length. The Roux limb is retrocolic, retrogastric, and the gastrojejunal anastomosis is performed using the linear stapler to create the posterior wall and a running suture to close the anterior wall. Main outcome measures included conversion rates, complications and weight loss at one year.

The weight reduction in the first 15 cases followed for more than one year was 67% of excess body weight. There were three conversions in this series (10.3%). One due to accidental perforation of the stomach and two due to inadequate exposure due to an enlarged left hepatic lobe. One patient developed an internal hernia with obstruction of the transverse mesocolic defect (3.4%). Four patients required postoperative blood transfusions due to postoperative hemorrhage. Two of these patients went on to develop anastomotic leaks requiring reoperation. One patient developed a marginal ulcer with perforation requiring reoperation 2-1/2 years after the initial procedure. There was no operative mortality in this series.

In June of 2000, Wittgrove and Clark published their results of 500 patients undergoing laparoscopic gastric bypass with 3-60 month follow-up.[3] Their technique consists of a 15 cc gastric pouch, 75 cm retrocolic, retrogastric Roux

Laparoscopic Bariatric Surgery: Techniques and Outcomes, edited by Eric J. DeMaria, Rifat Latifi and Harvey J. Sugerman. ©2002 Landes Bioscience.

limb, and a circular EEA stapled gastrojejunal anastomosis. Their main outcomes measured included pre- and postoperative co morbidities as well as weight loss over time. To date, this is the only publication documenting 60-month follow up following laparoscopic gastric bypass.

Their results showed 60% excess body weight loss six months after surgery and 77% one year after surgery, the best one-year result in the literature. Regarding longer-term weight loss, approximately 80% of patients lost and maintained 50% or more of their excess body weight 36-60 months postoperatively. Their 500 patients had a total of 1,752 comorbidities, which were reduced to 71, a 96% reduction following surgery. Weight loss was less in the diabetic patients compared to nondiabetic patients. However, 64/85 diabetics had elevated HgbA1C prior to surgery compared to only three postoperatively and all 39 type II diabetics on medication were off their medication postoperatively.

The anastomotic leak rate in this series was 2.2% (11/500). Nine of the 11 required reoperation, the majority of which were performed laparoscopically. The stomal stenosis rate was 1.6%. The wound infections were stratified into major, 0.8% and minor, 4.8%, the majority of which occurred at the trocar site used to introduce the circular EEA stapler.

Four out of 500 patients (0.8%) required re-exploration for hemorrhage. Three out of the four were successfully re-explored laparoscopically. Operating times early in the study averaged four hours and at the time of publication approached 90 minutes. The average length of stay was 2.5 days. There was no operative mortality.

In August of 2000, Nguyen et al published their results of 35 patients undergoing laparoscopic gastric bypass.[4] This was the first comprehensive study comparing laparoscopic and open gastric bypass patients. The data was collected prospectively in the laparoscopic group and compared to 35 retrospectively matched patients who underwent open gastric bypass. A 15 to 20 cc gastric pouch was created and the gastrojejunal anastomosis was performed using a circular 21mm EEA stapled anastomosis, passing the anvil transorally. The Roux limb was measured 75 cm in length (150 cm for BMI >50). Main outcome measures included operative time, blood loss, hospital stay, complications and weight loss.

Seventeen of 35 patients were available and evaluated for follow-up one year after laparoscopic gastric bypass. The mean percentage excess body weight loss at one year was 69%. Estimated blood loss was 135 cc and mean operating room time was 246 minutes. Mean length of stay was four days.

Two patients developed postoperative hemorrhage requiring transfusion. One patient (2.8%) developed a bowel obstruction secondary to narrowing of the jejuno-jejunostomy, which required laparoscopic revision. One patient developed respiratory failure requiring more than 72 hours of ventilatory support. Seven of 35 patients (20%) developed anastomotic strictures requiring endoscopic balloon dilatation. None of the patients developed venous thrombosis or pulmonary embolism. There were no conversions in this series. There were no anastomotic leaks in this series and no operative mortality.

In September of 2000, Higa et al published their results of 400 patients undergoing laparoscopic gastric bypass.[5] Their technique consisted of a 20 cc gastric pouch and a 100 cm retrocolic, antegastric Roux limb (150 cm for BMI >50). The gastrojejunal anastomosis is performed using a standard 2-layer hand sewn anastomosis. Main

11

outcome measures included weight loss, complication, length of stay, conversion rates and operative times.

Mean excess weight loss was 69% in 58 patients followed one year after surgery. This ranged from 37% of excess weight loss in a patient with a BMI >70 to 79% in 12 patients with an initial mean BMI less than 40.

In this series there have been no leaks, no deep venous thrombosis, and no operative mortality. There was a 5.25% incidence of stomal stenosis requiring balloon dilation. There were no trocar site hernias but 14 patients developed internal hernias, 13 of which were through the transverse mesocolon and one through the small bowel mesentery. Twelve of the 14 were successfully reoperated on laparoscopically. This high rate was attributed to the initial use of absorbable suture at these sites. They have since changed to a permanent suture for the closure of these defects with a current internal hernia rate of less than 1%. Four patients (1%) developed marginal ulcers, one of which required operative repair for a perforation. Operative time was 150-230 minutes earlier in their experience but at the time of publication approached 60-90 minutes as they improved their operative technique. Comorbidities and subsequent improvement following surgery was not documented in this paper.

In October of 2000, Schauer et al published 1-31 month follow-up results of 275 patients undergoing laparoscopic gastric bypass at the University of Pittsburgh.[6] Their technique consists of a 15 cc gastric pouch and a 75 cm retrocolic, retrogastric Roux limb (150 cm for BMI >50). The first 150 cases were performed using a circular EEA stapled gastrojejunal anastomosis. The final 125 consisted of an end to side gastrojejunostomy technique using an Endo GIA stapler. To date, this paper provides the most comprehensive follow-up of patients undergoing laparoscopic gastric bypass, detailing intraoperative complications, LOS, early and late postoperative complications and as well as weight loss and change in co-morbidities.

In this series, 43% of patients underwent a short limb bypass and 56% underwent a long limb bypass. One hundred and one patients had a mean excess weight loss of 68.8% one year after surgery. Nineteen patients had 83.2% excess weight loss two years after surgery and five patients maintained 76.7% excess weight loss 30 months after surgery.

Mean length of stay was 3.6 days with a range of 1-84 days. The mean operating room time was 260 minutes and mean operative blood loss was 115 cc with a conversion rate was 1.1%. A total of 12 gastrointestinal leaks occurred (4.4%), four of which were treated operatively (two laparoscopic and two open). One patient died of a fatal pulmonary embolus resulting in an operative mortality of 0.4%. Wound infections were 22% in the first 50 patients primarily at the trocar site where the EEA stapler is used. This dropped to 1.5% when the Endo GIA stapler was used eliminating the need for a contaminated instrument to come in contact with the skin. There were four bowel obstructions in this study (1.5%), two were at the jejuno-jejunostomy, one due to internal herniation through the mesocolic window and one six-months postoperatively due to adhesions.

Outcomes

Percentage Excess Weight Loss

If laparoscopic gastric bypass is to supercede open gastric bypass as the new gold standard in bariatric surgery, then the morbidity of laparoscopic gastric bypass must compare favorably to its open counterpart. Furthermore, percentage of excess weight loss over time must be equal if not better than in the open gastric bypass patients. Because the laparoscopic gastric bypass technique is relatively new, long-term follow-up is not available at this time. However, all of the previously reviewed series reported at least 1-year follow-up. These results ranged from 67% to 77% of excess weight lost in patients followed for one year following laparoscopic gastric bypass (Tables 11.1 and 11.2). These results are comparable to many of the reported series in the open literature. Sugerman et al, in a prospective study evaluating open gastric bypass to vertical banded gastroplasty, showed percentage excess weight loss at one year to be 68% in the open gastric bypass group.[7] Brolin et al showed a 72% excess weight loss at one-year follow-up in 108 patients undergoing open gastric bypass and Yale showed a 71% excess weight loss at one-year follow-up in 126 patients undergoing open gastric bypass.[8,9] To date, it appears that laparoscopic gastric bypass is equivalent to open gastric bypass with regards to percentage of excess weight loss at one-year follow-up. Further follow-up is necessary to see if the long-term results will be equally good.

Comorbidities

Two papers focused on associated co-morbidities related to obesity and their subsequent improvement following laparoscopic gastric bypass. Wittgrove and Clark as well as Schauer documented significant improvement or resolution in a multitude of co-morbidities ranging from 83% to 100% (Table 11.3). These results are consistent with those reported by Brolin et al, in which there was a 96% improvement or resolution of co-morbidities in 146 patients undergoing open gastric bypass.[8] Both Wittgrove and Schauer showed a 99-100% improvement and/or resolution in NIDDM. Pories showed a 91% reduction of NIDDM in 165 patients undergoing open gastric bypass. Fourteen-year follow-up revealed 83% of these patients maintained normal blood glucose values and glycosylated hemoglobin.[10] Long-term resolution of NIDDM in patients undergoing laparoscopic gastric bypass remains to be seen.

Internal Hernias

Higa, in his series of 400 patients noted 14 bowel obstructions secondary to internal hernias (3.25%) (Table 11.4). Thirteen of these 14 were from Roux limb migration through the mesocolic defect and one was through the small bowel mesentery of the jejuno-jejunostomy. Twelve of these 14 were successfully re-explored laparoscopically. Higa had originally used absorbable suture to close the mesenteric defects but has since changed to a permanent suture and has seen the incidence of internal hernia drop to less than 1%. Lonroth reported an internal hernia rate of 3.4% in his series of 29 patients. It is hypothesized that the laparoscopic approach results in less intraoperative adhesions and hence, a theoretically increased chance of internal herniation. At the Medical College of Virginia, during an open gastric

Table 11.1. Demographics

Investigator	N	Study Period	Mean F/U	Age Range/Mean	Female/ Male N	Preop BMI Range/Mean
Higa	400	22	N/A	13-70/43	330/70	35-78/46
Schauer	275	32	9.4	17-68/42	225/50	35-68/48
Wittgrove	500	N/A	N/A	N/A	N/A	35->55/N/A
Nguyen	35	12	12	22-59/41	30/5	40-60/51
Lonroth	29	33	N/A	22-63/33	24/5	N/A /42

Table 11.2. Percent excess weight loss in patients followed one year after laparoscopic gastric bypass

Investigator	%	N	Roux limb length obese/ superobese (cm)
Higa	69	58	100/150
Schauer	69	101	75/150
Wittgrove/Clark	77	N/A	75/
Nguyen	69	26	75/150
Lonroth/Dalenback	67	15	N/R

Table 11.3. Percent reduction in comorbidities

	Wittgrove/Clark % resolved	Schauer % improved or resolved
GERD	98.5	96
Hypercholesterolemia	97	96
Hypertriglycidemia	99	86
Diabetes II	99	100
Glucose Intolerance	100	—
Stress Urinary Incontinence	97	83
Sleep Apnea	98	93
Hypertension	92	88
Arthritis	90	88

11

bypass, we traditionally use absorbable chromic suture while closing the potential defects of the small bowel mesentery as well as transverse mesocolon and Petersen's hernia, but use nonabsorbable suture during the laparoscopic approach.

Wound Infections

An advantage that laparoscopic gastric bypass appears to offer over open gastric bypass is a significantly lower incidence of postoperative wound infections. In a review of 162 patients undergoing open gastric bypass, Sugerman et al had a 15.8%

incidence of wound infections, 4.4% of these were considered severe enough to delay hospital discharge.[11] Yale had a 5.9% incidence of wound infections in 251 patients undergoing open gastric bypass, 1.9% of these were considered major.[9] Hall had a 3.3% incidence of wound infection in 120 patients undergoing open gastric bypass.[13] The highest documented incidence of wound infection in the laparoscopic literature is in Schauer's initial 50 patients that underwent an EEA stapled anastomosis, resulting in a wound infection rate of 22%.[6] This was felt to be secondary to the operative technique in which an abdominal port is dilated to allow passage of the circular EEA stapler. In doing so, the contaminated stapler comes in contact with the skin. This high rate of wound infection was dramatically dropped to 1.5% when the linear stapled technique was employed, eliminating skin contact with contaminated instruments.

Wittgrove and Clark also use the circular stapler to perform their gastrojejunal anastomosis. However, they rely on the principles of the percutaneous endoscopy gastrostomy (PEG) technique and introduce the stapler orally. Their major and minor wound infection rates were .8% and 4.8%, respectively (Table 11.4).

Higa performs a two-layered hand-sewn gastrojejunal anastomosis and in 400 patients had no reported wound infections. At the Medical College of Virginia, we have not had any wound infections, to date, in our series of laparoscopic gastric bypass patients performed using the linear stapled technique.

Trocar Site Hernias

Controversy exists as to which sized trocar sites need to be closed, if any, during laparoscopic procedures, including laparoscopic gastric bypass. In this collective review, Nguyen is the only author who routinely closes trocar sites over 5 mm in diameter. The remaining authors do not routinely close any sized trocar sites. Nguyen, in his series of 35 patients, reported no trocar site hernias with 12-month follow-up. The remaining authors do not routinely close their trocar sites and, collectively, one patient in 1204 (.08%) developed a trocar site hernia. At the Medical College of Virginia, we do not routinely close our trocar sites and, to date, have not seen a trocar site hernia. Regardless of whether one chooses to close the trocar sites, it is clear that the incidence of abdominal wall hernias is exceedingly rare when compared to the open gastric bypass literature, where the incidence ranges from 2% to 21%.[9,12,13] Clearly, the dramatic reduction in frequency of abdominal wall hernias is a significant advantage that the laparoscopic approach has over the traditional open technique.

Anastomotic Leaks

The reported gastrojejunal anastomotic leak rates following open gastric bypass range from 0% to 5.6%, with most series quoting a less than 2% leak rate.[12-15] The reported anastomotic leak rate following laparoscopic gastric bypass in this series ranges from 0 in Higa's series of 400 patients to 6.9% in Lonroth's series of only 29 patients. The collective leak rate in the laparoscopic series is 25/1339 (2.0%). When discussing reported leak rates, it is important to define the term "leak" prior to interpreting the literature. Schauer divided his leaks into "clinical" and "subclinical". Clinical leaks were those that presented with peritonitis or an abscess. Four patients presented in such a fashion for a clinical leak rate of 1.5%. Eight patients

11

Table 11.4. Complications

	N	Conv. %	GJ Leak %	Mortality %	Internal Hernia %	Marginal Ulcer %	DVT/ PE%	Stomal Stenosis	Bleeding %	Bowel Obst. %	Trocar Site Hernias (%)	Wound Infections (%)	Symptomatic Cholelithiasis (%)	Type of Anastomosis
Higa	400	3	0	0	3.25	1	0/0	5.25	.25	3.25	0	N/A	2.2	Hand-Sewn
Schauer	275	1.1	4.4	0.4	0.4	0.7	0.7/0.4	4.7	3.3	1.5	0.3	8.7	1.5	150EEA 125 linear
Wittgrove	500	N/A	2.2	N/A	N/A	N/A	N/A	1.6	0.8	0.6	N/A	5.6	N/A	EEA
Nguyen	35	0	0	0	N/A	N/A	0/0	20	5.7	2.8	N/A	2.8	N/A	EEA
Lonroth	29	10.3	6.9	0	3.4	3.4	N/A	N/A	13.7	3.4	N/A	N/A	N/A	Linear staple

11

(2.9%) had subclinical leaks. One was identified on a routine postoperative UGI, which was repaired laparoscopically. One was identified by bile coming out of the drain on postoperative day 1 and repaired laparoscopically. Six patients were identified on postoperative day 8-10 by cloudy fluid in the Jackson-Pratt drain. All were managed with a combination of NPO for 1-2 weeks with intravenous fluids or a laparoscopically placed gastrostomy tube in the excluded stomach. Schauer's practice is to remove the drain on postoperative day 10. At the Medical College of Virginia, with over 175 laparoscopic gastric bypasses, we have only seen one leak present over one week after the surgery. Although the presence of a drain may prevent the need for a second operation in the face of an anastomotic leak, one must consider whether closed suction actually promotes a leak if the drain is left in place for an extended period of time. Schauer changed from an EEA stapled anastomosis to a linear stapled anastomosis for the last 125 cases in his series. He does not fractionate the leak rates based on which anastomosis was performed.

At this time, there are three recognized techniques for performing the gastrojejunal anastomosis; hand-sewn, EEA circular stapled, and linear stapled. Clearly the hand-sewn requires the highest level of laparoscopic skill and is reserved for that subset of advanced laparoscopists comfortable with intracorporeal suturing techniques. The EEA stapled anastomosis is the original technique as described by Wittgrove and Clark and probably is the most widely utilized of the three. It requires oral placement of the anvil, which is associated with its own set of complications, or it must be passed through a dilated trocar site in order to accommodate the larger diameter of the circular stapler. At the Medical College of Virginia, during a laparoscopic gastric bypass we perform a linear stapled anastomosis with a posterior and anterior row of oversewn running nonabsorbable suture. At this time, it appears that the leak rate is higher with laparoscopic gastric bypass and is probably, in part, due to the steep learning curve associated with laparoscopic gastric bypass.

When discussing anastomotic leaks in the context of laparoscopic gastric bypass surgery, one must not overlook the possibility of a leak occurring at the jejuno-jejunostomy. Review of this collective series shows that there was only one leak from the jejuno-jejunostomy for an incidence of 0.08%.

Conclusions

Review of the available literature shows many comparable results between the laparoscopic and open techniques. At this time, it appears that laparoscopic gastric bypass offers equivalent weight loss to open gastric bypass at one-year follow-up. The advantages of the laparoscopic approach appear to be a significant reduction in incidence of wound infections and abdominal wall hernias, possibly at the expense of a somewhat higher anastomotic leak rate. Further studies are in progress in order to provide enough data to establish laparoscopic gastric bypass as the new gold standard in bariatric surgery.

Selected Readings

1. Wittgrove A, Clark G, Tremblay L. Laparoscopic gastric bypass, roux-en-y: Preliminary report of five cases. Obes Surg 1994; 4:353-357.
2. Lonroth H, Dalenback J. Other laparoscopic bariatric procedures. World J of Surg 1998; 22:964-968.

3.   Wittgrove A, Clark G. Laparoscopic gastric bypass, roux en-y-500 patients: technique and results, with 3-60 month follow-up. Obesity Surg 2000; 10:233-239.

4.   Nguyen N, Ho H, Palmer L, et al. A comparison study of laparoscopic versus open gastric bypass for morbid obesity. J Am Coll Surg 2000; 191:149-157.

5.   Higa K, Boone K, Ho T, et al. Laparoscopic roux-en-y gastric bypass for morbid obesity. Arch Surg 2000; 135:1029-1034.

6.   Schauer P, Ikramuddin S, Gourash W, et al. Outcomes after laparoscopic gastric bypass for morbid obesity. Ann Surg 2000; 232(4):515-529.

7.   Sugerman H, Starkey J, Birkenhauer R. A randomized prospective trial of gastric bypass versus vertical banded gastroplasty for morbid obesity and their effects on sweets versus non-sweets eaters. Ann Surg 1987; 205(6):613-624.

8.   Brolin R, Robertson L, Kenler H, et al. Weight loss and dietary intake after vertical banded gastroplasty and roux-en-y gastric bypass. Ann Surg 1994; 220:782-90.

9.   Yale C. Gastric surgery for morbid obesity: Complications and long-term weight control. Arch Surg 1989; 124:941-946.

10. Pories W, Swanson M, MacDonald K, et al. Who would have thought it? An operation proves to be the most effective therapy for adult-onset diabetes mellitus. Ann Surg 1995; 222:339-50.

11. Sugerman H, Londrey G, Kellum J, et al. Weight loss with vertical banded gastroplasty and roux-en-y gastric bypass for morbid obesity with selective versus random assignment. Am J Surg 1989: 157:93-102.

12. Sugerman H, Kellum J, Engle K, et al. Gastric bypass for treating severe obesity. Am J Clin Nutr 1992; 55:560S-66S.

13. Hall J, Watts J, O'Brien P, et al. Gastric surgery for morbid obesity. The Adelaide study. Ann Surg 1990; 211:419-427.

14. Griffen O, Bivins B, Bell R, et al. Gastric bypass for morbid obesity. World J of Surg 1981; 5:817-822.

15. MacLean L, Rhode B, Sampalis J, et al. Results of the surgical treatment of obesity. Am J Surg 1993; 165:155-162.

11

Postoperative Management and Complications after Laparoscopic Roux-en-Y Gastric Bypass

John M. Kellum

Since the advent of laparoscopic Roux-en-Y gastric bypass by Wittgrove et al, in 1993, the operation has generally been accorded high marks as to feasibility, when performed by skilled laparoscopic surgeons, and as to weight loss, when compared with the original operation performed using conventional open surgical techniques. Early series suggest that some improvement in the risk for wound complications (infection and hernia) may be possible; on the other hand, there may be a higher risk of gastrojejunal anastomotic leak and stomal stenosis. In addition, the method of gastrojejunal anastomosis continues to undergo revision, based on a widespread dissatisfaction with the original technique, described by Wittgrove et al, in which an EEA™ (Ethicon Division of Johnson and Johnson, New Brunswich, NJ) 21 mm French-guage anvil must be pulled per orum down the esophagus, using an endoscopically positioned guide wire.

Postperative Management

Postoperative care following laparoscopy is similar to that after open Roux-en-Y gastric bypass. In general, postoperative pain is less so that respiratory therapy and ambulation can be pushed even more aggressively.

Perioperative low molecular weight heparin, in the form of enoxaparin 40 mg is given daily intramuscularly and is continued after the preoperative dose. Intermittent machine-driven compression stockings are used continuously intra- and postoperatively.

As in open gastric bypass, the patient's pain pattern, temperature and pulse rate are closely monitored. Special attention is paid to complaints of back or shoulder pain which can herald the occurrence of an anastomotic leak. A pulse rate greater than 125 is assumed as evidence of such a leak unless explained otherwise.

Our practice has been to obtain an upper gastrointestinal contrast radiographic study on the first postoperative day. The radiologist administers a water soluble solution, such as Gastrografin®, in order to exclude a leak at the gastrojejunostomy or the jejuno-jejunostomy. If no leak is demonstrated, barium sulfate may be administered to better define anastomotic calibre and gastrointestinal motility.

If this study is normal, the patient is permitted clear liquids up to 60 ml per hour. On the second postoperative day, pureed food, containing no added sugar, may be allowed. If the patient tolerates at least liquids, the Jackson-Pratt® drain is

Laparoscopic Bariatric Surgery: Techniques and Outcomes, edited by Eric J. DeMaria, Rifat Latifi and Harvey J. Sugerman. ©2002 Landes Bioscience.

removed and the patient is discharged home. Schauer's practice at the University of Pittsburgh has been to leave the drain in place for a full 10 days after surgery.

The patient is instructed to continue the pureed diet for a full 30 days to allow healing of the staple line in the gastric pouch. They are encouraged to eat at least 50 g protein per day. After 30 days, the patient is permitted regular food but instructed not to add sugar or drink liquids with sugar. Intolerance to red meat, and even poultry, is common during the first six to nine months after surgery. Many patients achieve their protein quota by eating fish, daily products, yogurt and protein supplements during this phase.

The daily supplement regimen includes 0.5 mg of vitamin B_{12}, 1.2 g of calcium, 650 mg of ferrous sulfate (in menstruating females) and a multivitamin tablet. Patients are strongly encouraged to develop a daily exercise program to maximize their resting metabolic rates. Many patients can achieve a daily three mile walk over the first few postoperative months. They are also discouraged from eating high-fat, high-calorie "junk foods," such as potato chips and popcorn.

Patients are seen back at seven to ten days postoperatively for surgical and nutritional follow-up. Nutritional follow-up is generally scheduled at quarterly intervals during the first postoperative year and annually thereafter. Before seeing his or her surgeon, the patient is seen by a registered dietician, who estimates calorie and protein intake, as well as compliance with vitamin and mineral therapy. Many also choose to join support groups in order to share menus and reinforce good exercise habits in groups.

Complication Rate after Open Gastric Bypass

In our own series of 672 open gastric bypass patients complications included a 30-day operative mortality rate of 0.4%. In 162 patients followed at least 5 years, there was a 1.2% incidence of anastomotic leak with peritonitis, 2% deep venous thrombosis and/or pulmonary embolism, 4.4% deep wound infection, 11.4% superficial wound infection or seroma, 1% incidence of gastric staple line disruption, 14.6% stomal stenosis requiring endoscopic balloon dilitation, 13.3% marginal ulcer, 19% incisional hernia and 10% symptomatic gallbladder disease requiring cholecystectomy.

Conversion from Laparoscopic to Open Roux-en-Y Gastric Bypass

Most authors have reported that conversion rates contract with increasing experience with the laparoscopic approach. Schauer et al reported a 1.1% conversion rate in a series of 275 patients, however, most occurred in their first 50 patients. They and others have noted the importance of proper trocar placement and the availability of long endoscopic instruments.

Mortality after Laparoscopic Roux-en-Y Gastric Bypass

In the published series of this operation which now numbers greater than 3000 patients, there are only two early postoperative deaths. These were fatal pulmonary emboli reported one each by the groups in Pittsburgh and Fresno. We are loathe to claim a lower mortality rate for this operation, since we encountered a higher rate of anastomotic leak in our early experience. It is likely that any lower mortality rate is associated with patient selection. In other words, patients who are super-obese (>50 kg/m^2), and who, because of their massive size, have life-threatening complications of obesity, such as severe obstructive sleep apnea, or who have had prior extensive upper abdominal surgery are less likely to be offered laparoscopic Roux-en-Y gastric bypass.

Anastomotic Complications

We and others encountered a higher rate of postoperative anastomotic leak and stenosis early in our experience with the operation. However, the routine use of a drain in proximity to the gastrojejunostomy appears to lessen the likelihood of generalized peritonitis, so that many of these leaks were contained and could be managed nonoperatively. Complications, such as perforated pharynx or esophagus, have been reported as a consequence of the passage of the EEA™ anvil, using an endoscopically positioned guide wire by mouth. Because of the risk of such an injury and because the transabdominal passage of the circular stapling device requires a larger trocar size, we and others have converted to using a linear stapler with anastomoses done entirely intra-abdominally.

Wittgrove et al reported a 5% leak rate in their first 75 patients. In a later series of 500 patients, however, the leak rate had been reduced to 1% in the most recent 200 patients. Schauer et al reported 12 (4.4%) rate of anastomotic leak, but of these, eight (2.9%) were asymptomatic or contained. Higa et al reported no clinical anastomotic leaks in 1040 patients who had a two-layer, hand-sewn anastomosis performed with absorbable suture material. In our own experience, it appears that an early higher anastomotic leak rate has been improved by using a linear stapling technique with closure of the stapling defect with a hand-sewn, two-layer running absorbable suture.

Stenosis may occur either at the gastrojejunostomy or the jejuno-jejunostomy. Our incidence of problems at the latter anastomosis has been reduced with experience and with the use of a single firing of the Endo-GIA -6.0 cm, 2.5 mm stapling cartridge (U.S. Surgical division of Tyco, International, Norwalk Connecticut). Since prior experience with the circular stapler in esophagectomy, as reported by Skinner et al, has indicated a higher rate of anastomotic stenosis but a lower rate of anastomotic leak, it is not surprising that the stenosis rate for the gastrojejunostomy remains significant. On the other hand, endoscopic dilitation with the TTS balloon series of either a circular, linear-stapled or handsewn gastrojejunostomy has a high rate of success. Schauer et al used both the circular and linear stapler for perform gastrojejunostomies in 275 patients and reported a 4.7% rate of stomal stenosis requiring endoscopic balloon dilatation, all of which were successful. Higa et al reported a 4.9% rate of stomal stenosis, all of which were successfully dilated endospically, in 1040 patients with handsewn anastomosis.

Reported marginal ulcer rates for laparoscopic Roux-en-Y gastric bypass appear to be lower than those reported for the open operation. Schauer et al reported only 2 in his series of 275, both of which resolved on proton pump inhibitor therapy. Higa et al reported 14 patients (1.4%) with ulcers in a series of 1040 patients. Two presented with perforation. He noted that most were associated with patients taking NSAID's despite precautions against doing so without protection. Most marginal ulcers, whether after open or laparoscopic Roux-en-Y gastric bypass, can be successfully

12

treated with intensive proton pump inhibitor therapy, sucralfate and withdrawal from NSAID's.

Acute Gastric Dilitation

This complication is caused by early postoperative partial obstruction at the jejuno-jejunostomy. It can result in staple line blowout or anastomotic leak. It is heralded by hiccoughs, shoulder pain and hypotension. A plain abdominal radiograph will usually confirm the diagnosis. The complication is usually preventable by careful attention to technique, when constructing the distal anastomosis and by avoiding distal segments of jejunum with a small diameter. Laparoscopic decompression of the distal, bypassed stomach is the treatment of choice.

Thromboembolic Complications

There is little reason to think that such complications will be different because of the approach. Most of the risk derives from patient factors, such as obesity, increased intra-abdominal pressure and a history of prior thromboembolic events. While theoretically, insufflation of CO_2 gas under pressure might further inhibit the return of venous blood from the lower extremities, the actual reported thromboembolic rate of complications has not been higher with the laparoscopic approach. Our current critical care pathway for laparoscopic Roux-en-Y gastric bypass includes routine use of pre- and postoperative low molecular weight heparin (enoxaparin 40 mg, IV daily), while the patient is in the hospital, and the intra- and postoperative use of intermittent compression boots.

Wound Complications

In various series of laparoscopic gastric bypass there appears to be a reduced incidence of wound infection. Wittgrove and Clark reported a 5% incidence of wound infection (most of which they describe as minor and involving only one trocar site) in 500 patients. Nguyen et al reported no severe wound infections in 35 laparoscopic patients, as compared to 2 in 35 open gastric bypass patients. Higa reported only one wound infection in over 1000 patients. Schauer et al reported a 4.7% incidence of purulent wound infections in 275 laparoscopic patients. Several factors may explain this possible reduction in the incidence of wound infection. Firstly, the overall abdominal wall wound area is vastly reduced by the laparoscopic approach. Secondly, the use of endoscopic linear stapling devices which both staple and divide the intestines and stomach limit the time exposure of the wound to lumenal contents. Since the staplers are withdrawn within trocars, there is no direct exposure of the wounds to contaminated instruments or tissue.

The highest potential advantage of the laparoscopic approach, however, appears to be in a significant reduction in the incidence of incisional hernia. Wittgrove and Clark reported no incisional hernias in 500 patients. Higa reported only 3 trocar hernias (0.3%) in 1040 patients. Schauer et al reported a 0.3% incidence of incisional hernia in his series of 275 patients. In our own series at the Medical College of Virginia at Virginia Commonwealth University we have had two incisional hernias since we went to a totally laparoscopic, totally transabdominal approach. This series is now over 300 patients. In other words, since we began using a linear stapling technique for our gastrojejunostomy, our largest trocar wound has been of the 12 mm size.

Cholelithiasis

Sugerman et al (1995) reported that the routine use of prophylactic ursodiol (600 mg/day) reduced the risk of postoperative gallstone formation from 32% to 2% after open Roux-en-Y gastric bypass. Higa et al reported only a 1.4% incidence of clinically symptomatic gallstones in his large series of laparoscopic Roux-en-Y gastric bypasses. These authors noted that the majority of these patients failed to take their prescribed ursodiol.

Staple Line Disruption

Higa et al reported an episodic "run" of failed staple-lines, associated with a change in "manufacturing specifications." He noted that this has been rare with the most recent endoscopic stapling instruments. We have noted a few patients with fistulas between proximal and gastric pouches which resolved spontaneously. Careful application of the endoscopic stapler with a slow, smooth hand closure of the device is important. When dealing with thickened, scarred stomach or intestine, we recommend careful oversewing of sutures lines with running absorbable suture.

Internal Hernia

This complication is potentially very dangerous, because it can lead to strangulation necrosis of large segments of small intestine. We and others have pointed out the importance of closing each of three defects: the mesenteric defect under the jejuno-jejunostomy, the defect in the transverse mesocolon and the so-called "Petersen" defect between the small intestinal mesentery of the Roux-en-Y limb and the underlying mesocolon. As Higa et al have noted, it is likely that this problem may be even higher with the laparoscopic approach since fewer adhesions are created. Reporting a 2.5% incidence of internal hernia, requiring reoperation, they recommended careful closure of all defects using running non-absorbable suture.

Diagnosis of this complication may be difficult since the patient may report only intermittent episodes of obstruction which evade radiographic diagnosis. It should be suspected in any patient having such symptoms. A widened GIA staple line in an unusual location, such as the left upper quadrant of the abdomen, should raise the index of suspicion. Either immediate computerized tomography of the abdomen or an upper intestinal contrast radiograph, when the patient is actually having the symptoms can confirm the diagnosis. If these studies do not confirm the diagnosis but symptoms continue, exploratory celiotomy may be indicated.

Surgical exploration may be done laparoscopically in many patients. The defects listed above can all be closed laparoscopically. However, if a strangulation obstruction is found, conversion to open surgery may be necessary.

Summary

There is an extensive list of potential complications for this difficult surgical operation. In open gastric bypass, nearly 35% of patients suffer some type of complication. There is evidence, however, that certain complications, such as wound infection and incisional hernia, will be significantly lower with the laparoscopic approach. It is also becoming clear that the problem of anastomotic leak will become less common with evolving. laparoscopic techniques. Only surgeons thoroughly committed to the long-term care of bariatric surgical patients should attempt this operation. These patients tend to be demanding and have a relatively high incidence of chronic medical and surgical problems which require a surgeon dedicated to the field of bariatric surgery.

Selected Readings

1.　Higa KD, Boone KB, Ho T. Complications of the laparoscopic roux-en-Y gastric bypass: 1,040 patients-what have we learned? Obes Surg 2000; 10:509-13.

2.　Kellum JM, DeMaria EJ, Sugerman HJ. The surgical treatment of morbid obesity. Curr Prob Surg 1998; 35:796-851.

3.　Nguyen NT, Ho HS, Palmer LS, et al. A comparison study of laparoscopic versus open gastric bypass for morbid obesity. J Am Coll Surg 2000; 191:149-55.

4.　Nguyen NT, Wolfe BM. Hypopharyngeal perforation during laparoscopic Roux-en-Y gastric bypass. Obes Surg 2000; 10:64-7.

5.　Serra C, Baltasar A, Bou R, et al. Internal hernias and gastric perforation after a laparoscopic gastric bypass. Obes Surg 1999; 9:546-9.

6.　Schauer PR, Ikramuddin S, Gourash W, et al. Outcomes after laparoscopic Roux-en-Y gastric bypass for morbid obesity. Ann Surg 2000; 232:515-29.

7.　Schweitzer MA, Broderick TJ, DeMaria EJ, et al. Laparoscopic-assisted Roux-en-Y gastric bypass. J Laparoendosc Adv Surg Tech A 1999; 9:449-53.

8.　Sugerman HJ, Brewer WH, Shiffman ML, et al. A multicenter, placebo-controlled, randomized, double-blind, prospective trial of prophylactic ursodiol for the prevention of gallstone formation following gastric-bypass-induced rapid weight loss. Am J Surg 1995; 169:91-7.

9.　Wittgrove AC, Clark GW, Tremblay LJ. Laparoscopic gastric bypass, Roux-en-Y: Preliminary report of five cases. Obes Surg 1994; 4:353-57.

10.　Wittgrove AC, Clark GW. Laparoscopic gastric bypass, Roux-en-Y- 500 patients: Technique and results, with 3-60 month follow-up. Obes Surg 2000; 10:233-9.

12

Index

W